FRESHWATER FISH PARASITES

FRESHWATER FISH PARASITES

by

Prathap. C

Anupama S

Wordpen Academics

Publishing Science That Matter

ISBN: 978-93-6084-700-5 (Hardback)

Published by : **WORDPEN ACADEMICS**
4736/23, Ansari Road, Darya Ganj,
New Delhi – 110 002
Phone: +91-011-35004336
E-mail: info@wordpenacademics.com

Printed at **:** **Replika Press Pvt. Ltd.**

Exclusive Distributor: ***Bio-Green Books, New Delhi***

PREFACE

Freshwater Fish Parasites is an in-depth exploration of the diverse range of parasitic organisms that affect freshwater fish, their impact on fish health, and the methods used for their detection, management, and control. This book is designed for students, researchers, and professionals in the fields of aquaculture, fisheries science, parasitology, and veterinary medicine, providing a thorough understanding of the ecological and economic importance of freshwater fish parasites.

Freshwater fish are vital to global food security, supporting both wild fisheries and aquaculture industries. However, they are also susceptible to a wide variety of parasitic infections that can lead to severe economic losses and environmental challenges. The study of these parasites is essential not only for the health of the fish but also for the sustainability of the fisheries industry and the preservation of aquatic ecosystems.

This book begins with a foundational overview of the types of parasitic organisms that affect freshwater fish, categorizing them into protozoa, helminths (worms), and ectoparasites (external parasites). Each group is examined in detail, with a focus on their biology, life cycles, and modes of transmission. The book also delves into the pathology of these infections, exploring how parasites affect the physiology and behavior of fish, leading to reduced growth rates, compromised immune systems, and in severe cases, fish mortality.

In addition to understanding the biology and pathology of freshwater fish parasites, Freshwater Fish Parasites addresses the various diagnostic techniques used in parasitology. It covers the use of microscopy, molecular methods, and histopathology for parasite identification and detection. Accurate diagnosis is crucial for effective treatment and management, and the book provides guidance on how to recognize parasitic infections in both wild and farmed fish populations.

A central theme of this book is the strategies used to manage and control parasitic infections in freshwater fish. The book explores traditional methods of parasite control, such as the use of anti-parasitic drugs and treatments, as well as integrated approaches that combine good husbandry practices, environmental management, and the use of biological control agents. Furthermore, it discusses emerging techniques such as genetic selection for parasite resistance, vaccine development, and the role of probiotics in enhancing fish health and immunity.

The book also highlights the role of environmental factors in the prevalence and transmission of parasitic diseases. Changes in water quality, temperature, and habitat conditions can all influence the dynamics of parasitic infections, making environmental management a key component of disease prevention. Special attention is given to the impact of climate change on freshwater ecosystems and the potential for altering parasite distributions and infection patterns.

Throughout the book, case studies and real-world examples illustrate the practical challenges and successes of managing freshwater fish parasites in various settings, from small-scale aquaculture farms to large commercial operations. These examples provide insights into the complexities of parasite management and offer guidance on how to tackle these challenges in different geographical and ecological contexts.

Freshwater Fish Parasites also emphasizes the importance of education and awareness in preventing the spread of parasitic infections. The book provides recommendations for best practices in fish farming, biosecurity protocols, and public health initiatives that aim to reduce the risk of parasitic diseases in both aquaculture and wild fish populations.

In conclusion, this book serves as a comprehensive resource on the study of freshwater fish parasites, offering both theoretical insights and practical approaches to parasite management. It is our hope that Freshwater Fish Parasites will contribute to the advancement of research in parasitology, provide valuable guidance for industry professionals, and help improve the health and sustainability of freshwater fish populations worldwide.

We would like to express our gratitude to the contributors, researchers, and practitioners who have shared their expertise and research to make this book a comprehensive reference on this important subject.

Prathap. C
Anupama S

CONTENTS

PREFACE V

1. INTRODUCTION 1
   - Habitat Destruction 1
   - The rainbow trout is an invasive species in many ecosystems 3
   - Sustainable Fisheries 7
   - Traditional management of fisheries 8
   - Environmental impact of fishing 15
   - Fishing down the food web 17
   - Destructive fishing practices 21
2. FRESH WATER FISH CULTURE 23
3. MARINE FISHERIES ISSUES 41
   - Techniques 41
   - Marine Protected Area 42
   - Marine Pollution 54
   - Mercury in Fish 55
   - Shark Finning 58
4. MARINE FISHERIES ORGANIZATION 63
   - National Marine Fisheries Service 63
   - Bangladesh Marine Fisheries Academy 65
   - Gulf States Marine Fisheries Commission 70
   - Marine Stewardship Council 71
   - SeaChoice 73
   - Seafood Watch 74
   - Oceana 76
   - WorldFish Center 79
   - Hotspot Ecosystem Research and Man's Impact On European Seas 82
   - PROFISH 90

- International Seafood Sustainability Foundation 91
- Sea Shepherd Conservation Society 91

5. INLAND FISHERIES 97

- What inland fisheries contribute to economic and social development ? 98
- Exotic Fishes 101
- Geographical Limitation of Inland Fisheries 103
- The Nile tilapia as the benchmark species 104
- Continental mask based on FAO inland Fisheries Statistical Areas. 106
- Rsults 110
- How are enhancement prospects affected by climate? 113
- Enhancements in small water bodies 115
- How are enhancement prospects affected by availability of natural fertilisers? 117
- Discussion 118
- Approaches to the estimation of enhancement potential 120
- Conclusions 121

6. METHOD OF INLAND FISHERIES 123

- Principles of Designing Inland Fish Farms 133
- Purpose and Dimensioning of Fish Farm Structures 137
- Fish Pond Arragements 140
- Dike Protection 144
- Structures 146
- Inland Capture Fisheries 148
- Comparative evaluation of inland fisheries 153
- Inland fisheries in low-income food-deficit countries 156
- Status and trends of inland capture fisheries production 157
- Trends in inland capture fisheries production 1986-2001 158
- Status 2000 159
- Trends in Organism Identification 160
- The most important species and groups in inland fisheries production 161

7. MANAGEMENT OF INLAND FISHERIES 165

- Role of FAO Inland Fishery Bodies 166

- Data Requirements for Inland Fisheries Management 170
- Complexity of Information: 174
- Information for Specific Purposes 178
- Conclusion 179

8. ANTIBIOTICS IN SALMONID AQUACULTURE 183

- Environmental impacts of antibiotics in fish farming 184
- Retarded decomposition of organic material 185
- Excessive Antibiotics 186
- Impacts of salmonid farming on marine coastal ecosystems 190
- Getting a Franken-Leg Up in the Annual Salmon Run 196

9. ECOSYSTEM BASED FISHERIES MANAGEMENT 199

- Background 199
- Institutional Foundations 209
- Ecosystem Approach Fisheries Principles 216
- Is Ecosystem Approach a Fusion or Collision? 223

10. ENVIRONMENTAL IMPACT OF FISHING 229

- Effects on habitat 229
- Extensive aquaculture 232
- Fisheries Impact on The Ecosystem 241
- Fisheries Versus other Impacts 246

BIBLIOGRAPHY 249

INDEX 251

1

INTRODUCTION

Freshwater fish are those that spend some or all of their lives in fresh water, such as rivers and lakes, with a salinity of less than 1.05%. These environments differ from marine conditions in many ways, especially the difference in levels of salinity. To survive fresh water, the fish need a range of physiological adaptations.

41.24% of all known species of fish are found in fresh water. This is primarily due to the rapid speciation that the scattered habitats make possible. When dealing with ponds and lakes, one might use the same basic models of speciation as when studying island biogeography.

Freshwater fish differ physiologically from salt water fish in several respects. Their gills must be able to diffuse dissolved gases while keeping the salts in the body fluids inside. Their scales reduce water diffusion through the skin: freshwater fish that have lost too many scales will die. They also have well developed kidneys to reclaim salts from body fluids before excretion.

Many species of fish do reproduce in freshwater, but spend most of their adult lives in the sea. These are known as anadromous fish, and include, for instance, salmon, trout, sea lamprey and three-spined stickleback. Some other kinds of fish are, on the contrary, born in salt water, but live most of or parts of their adult lives in fresh water; for instance the eels. These are known as catadromous fish.

Species migrating between marine and fresh waters need adaptations for both environments; when in salt water they need to keep the bodily salt concentration on a level lower than the surroundings, and vice versa. Many species solve this problem by associating different habitats with different stages of life. Both eels, anadromous salmoniform fish and the sea lamprey have different tolerances in salinity in different stages of their lives.

HABITAT DESTRUCTION

Intentional anthropogenic reconstruction and rerouting of waterways impacts stream flow, water temperature, and more, impacting normal habitat functionality. Dams not only interrupt linear water flow and cause major geological channel shifts, but also limit the amount of water available to fishes in lakes, streams and rivers and have the potential to change the trophic structure because of these alterations of the habitat and the limitations to movement and connectivity.

Unnatural water flow below dams causes immense habitat degradation, reducing viable options for aquatic organisms. Upstream migration is hindered by the dam structure and can cause population declines as fishes don't have access to normal feeding and/or spawning grounds. Dams tend to affect upstream species richness, that is, the number of fish species in the ecological community. Additionally, dams can cause the isolation of fish populations, and the lack of connectivity creates possible problems for inbreeding and low genetic diversity. The loss of connectivity impacts the structure of community assemblies and increases the fragmentation of habitats, which can compound existing problems for vulnerable species.

Temperature alterations are another unintended consequence of dam and land use projects. Temperature is a very important part of aquatic ecosystem stability, and thus changes to stream and river water temperature can have large impacts on biotic communities. Many aquatic larvae use thermal cues to regulate their life cycles, mostly notably here, insects. Insects are a large part of most fish diets, so this can pose a great dietary problem. Temperature can cause changes in fish behavior and distribution habits as well by increasing their metabolic rates and thus their drive to spawn and feed.

Linear systems are more easily fragmented and connectivity in aquatic ecosystems is vital. Freshwater fishes are particularly vulnerable to habitat destruction because they reside in small bodies of water which are often very close to human activity and thus easily polluted by trash, chemicals, waste, and other agents which are harmful to freshwater habitats.

Land use changes cause major shifts in aquatic ecosystems. Deforestation can change the structure and sedimentary composition of streams, which changes the functionality of the habitat for many fish species and can reduce species richness, evenness, and diversity. Agriculture, mining, and basic infrastructural building can degrade freshwater habitats. Fertilizer runoffs can create excess nitrogen and phosphorus which feed massive algae blooms that block sunlight, limit water oxygenation, and make the habitat functionally unsustainable for aquatic species. Chemicals from mining and factories make their way into the soil and go into streams via runoff. More runoff makes its way into streams since paved roads, cement, and other basic infrastructure do not absorb materials, and all the harmful pollutants go directly into rivers and streams. Fish are very sensitive to changes in water pH, salinity, hardness, and temperature which can all be affected by runoff pollutants and indirect changes from land use.

Exotic Species

An exotic (or non-native) species is defined as a species that does not naturally occur in a certain area or ecosystem. This includes eggs and other biological material associated with the species. Non-native species are considered invasive if they cause ecological or economic injury.

The introduction of exotic fish species into ecosystems is a threat to many endemic populations. The native species struggle to survive alongside exotic species which decimate prey populations or outcompete indigenous fishes. High densities of exotic fish are negatively correlated with native species richness. Because the exotic species was suddenly thrown into a community instead of evolving alongside the other organisms, it doesn't have established predators, prey, parasites, etc. which other species do, and the exotic species thus has a fitness advantage over endemic organisms.

One such example is the destruction of the endemic cichlid population in Lake Victoria via the introduction of the predatory Nile perch (Lates niloticus). Although the exact time is unknown, in the 1950s the Ugandan Game and Fisheries Department covertly introduced the Nile perch into Lake Victoria, possibly to improve sport fishing and boost the fishery. In the 1980s, the Nile perch population saw a large increase which coincided with a great increase in the value of the fishery. This surge in Nile perch numbers restructured the lake's ecology. The endemic cichlid population, known to have around 500 species, was cut almost in half. By the 1990s, only three species of sport fish were left to support the once multispecies fishery, two of which were invasive. More recent research has suggested that remaining cichlids are recovering due to the recent surge in Nile perch commercial fishing, and the cichlids that are left have the greatest phenotypic plasticity and are able to react to environmental changes quickly.

THE RAINBOW TROUT IS AN INVASIVE SPECIES IN MANY ECOSYSTEMS

The introduction of the rainbow trout (Oncorhynchus mykiss) in the late 19th century resulted in the extinction of the yellowfin cutthroat trout (Oncorhynchus clarkii macdonaldi) found only in the Twin Lakes of Colorado, USA. The yellowfin cutthroat trout was discovered in 1889 and was recognized as a subspecies of the cutthroat trout (Oncorhynchus clarkii). The rainbow trout was introduced to Colorado in the 1880s. By 1903, the yellowfin cutthroat trout stopped being reported. It is now presumed extinct. The rainbow trout is invasive worldwide, and there are multiple efforts to remove them from their non-native ecosystems.

Both species are among the "100 of the World's Worst Invasive Alien Species," as determined by the IUCN Invasive Species Specialist Group based on their effect on anthropogenic activities, environmental biodiversity and their ability to act as a case study for important ecological issues.

Hybridization

Hybridization is defined as the mating of two genetically different species (interspecific hybridization). It is dangerous for native species to hybridize because hybrid phenotypes may have better fitness and outcompete the two parent species and / or other fishes in the ecosystem. This could irreversibly compromise the genetic identity of one or both of the parent species and even drive them to extinction if their range is limited.

The rainbow trout discussed above hybridized with the native greenback cutthroat trout (Oncorhynchus clarkii stomias), causing their local extinction in the Twin Lakes area of Colorado as their hybrid "cutbows" became more prevalent. The rainbow trout has been reported to hybridize with at least two other salmonid species. Additionally, the cichlids in Lake Victoria evolved over 700 unique species in only 150,000 years and are theorized to have done so via ancient hybridization events which led to speciation.

Some Parasites of Freshwater Fish

Some of the more obvious or commonly encountered parasites of freshwater fish are briefly described below. Further details of their life histories, effects on fish and methods of control are given in the references.

Most fish in the wild carry some parasites. These are sometimes obvious but more often are difficult to detect other than by specialist techniques and usually appear to have little effect on the host fish.

However, in times of stress, resistance of fish is often lowered and some parasites may greatly increase in abundance and affect the health of the fish. In this situation fish will often lose condition, making them more susceptible to predation, or may even die from the effects of the parasites.

Fish which are injured in some way, such as after an attack by a predator, may carry a wound which is then infected by a parasite.

Fungal Infections

Fungal infections are characterised by white growths, like cotton wool, on the body and fins of infected fish.

Usually fungus infects only fish which have suffered some wound or damage or are weakened or stressed from some other disease. The fungus is often identified as Saprolegnia, but several other species of fungus may also infect fish.

Protozoan (single-celled animal) parasites, such as Costia, Chilodonella, Triochodena and Ichthyophthirius, may infect fish, especially under aquarium or intensive culture conditions.

Protozoans cannot usually be identified except under a microscope, but disease symptoms, such as a grey cloudiness of the skins and fins, laboured respiration, emaciation and unusual swimming behaviour, may give some clue to the causative agent.

Ichthyophthirius infections are known commonly as "white spot" because the parasite causes the formation of small, white cysts on the skin and fins of infected fish.

Tapeworms and Flukes

Several cestodes (tapeworms) and trematodes (flukes) are parasites on freshwater fish.

One such tapeworm is Ligula whose life cycle consists of several stages which are parasitic in turn on crustaceans, fish and birds.

Fish infected with Ligula exhibit grossly distended abdomens. The contained worms are white, segmented and may be up to 200 mm long, often longer than the infected fish.

Heavily-infected fish are weak and slow-moving, making them an easy target for predators. Galaxiids (minnows), particularly the Common Galaxias, are often infected by Ligula.

Several larval trematodes infecting fish causing what is commonly known as "black spot" because of the characteristic, small (about 2 mm in diameter) dark brown or black spots which develop in the muscle and on the body, fins, gills and eyes of infected fish.

When the parasite infects the fish it forms a cyst within the host tissue. The cyst then becomes surrounded by pigment cells, giving it the characteristic dark colour. "Black spot" infections occur in several species of freshwater fish, but galaxiids appear to be particularly susceptible to infection, with some fish carrying hundreds of cysts on the body and fins.

The organism causing "black spot" has been previously identified as Neascus, but there are several species of trematodes which have larval stages which cause black spots; these species have yet to be identified. The adult trematode is generally found infecting fish-eating birds.

Eustrongyloides is a nematode worm, the larval form of which is usually found in the muscle and internal organs of trout, redfin (English perch) and some native fish.

Heavily-infected fish may be severely emaciated and have numerous lumps under the skin, each cavity containing a thin, coiled, red worm about 20-30 mm long. Fillets taken from infected fish may have several worms in them, but if the flesh is cooked properly the worms are easily removed and the flesh is quite safe to eat.

Eustrongyloides is spreading steadily through Victorian waters, lakes particularly, and this is due to the natural cycle of fish-eating birds consuming infected fish and transferring either live worms or worm eggs to other waters. There is no means of control available, and several waters in south-western Victoria now contain fish populations which are heavily infested with this nematode.

Leeches

Leeches are also parasitic on freshwater fish. They attach themselves to the skin with their sucker and feed on the blood of the host. They leave characteristic Y-shaped scars, which may be susceptible to infection by other parasites.

Freshwater mussels

The larvae (glochidia) of freshwater mussels are parasitic on fish. They are released into the water by adult mussels and, when a fish passes close enough to disturb them, the glochidia attach themselves to the skin or gills of the fish by means of their barbed valves. Irritated host tissue then grows and forms a cyst over each glochidium.

Development from glochidium to small mussel takes about 10 weeks, at which time the mussel bores through the cyst, leaves its host and settles to the substrate.

The presence of a glochidia infestation is indicated by numerous white or greyish "bladders" on the gills, skin and fins of the fish. Fish may be severely stress by the attachment of large numbers of glochidia, particularly when the infestation affects the gills and may greatly impair respiration.

Glochidia are able to affect most native species but are not known to affect introduced species.

Crustaceans

Several crustaceans are parasitic on fish. Lernaea, although commonly called anchor worm, is in fact a crustacean. When attached to a fish, the head is buried in the flesh causing an inflamed, red wound. Only the worm- like body and twin egg-sacs are exposed.

The presence of large numbers of these parasites may severely stress the host fish, and the large, deep wounds produced are often then infected by bacteria or fungi.

Many native fish, such as Murray cod, golden perch and Macquarie perch, are susceptible to Lernaea attack, as are introduced salmonids (brown trout, rainbow trout and chinook salmon) and cyprinids (carp, goldfish, tench, roach).

Argulus, the fish louse, is a flat, oval, crab-like organism, with eight swimming legs and a small, flattened tail.

It swims freely through the water until it finds a host, then attached itself with the aid of two large suckers and forces it sharp trunk-like mouth into the fish to feed on its blood.

Large numbers of Argulus may weaken host fish because of the amount of blood they withdraw. The small, red, attachment sites are also susceptible to fungal infection.

Argulus commonly infects cyprinids, especially if held under aquarium conditions.

SUSTAINABLE FISHERIES

Sustainability in fisheries combines theoretical disciplines, such as the population dynamics of fisheries, with practical strategies, such as avoiding overfishing through techniques such as individual fishing quotas, curtailing destructive and illegal fishing practices by lobbying for appropriate law and policy, setting up protected areas, restoring collapsed fisheries, incorporating all externalities involved in harvesting marine ecosystems into fishery economics, educating stakeholders and the wider public, and developing independent certification programs.

The primary concern around sustainability is that heavy fishing pressures, such as overexploitation and growth or recruitment overfishing, will result in the loss of significant potential yield; that stock structure will erode to the point where it loses diversity and resilience to environmental fluctuations; that ecosystems and their economic infrastructures will cycle between collapse and recovery; with each cycle less productive than its predecessor; and that changes will occur in the trophic balance.

Global wild fisheries are believed to have peaked and begun a decline, with valuable habitats, such as estuaries and coral reefs, in critical condition. Current aquaculture or farming of piscivorous fish, such as salmon, does not solve the problem because farmed piscivores are fed products from wild fish, such as forage fish. Salmon farming also has major negative impacts on wild salmon. Fish that occupy the higher trophic levels are less efficient sources of food energy.

Fishery ecosystems are an important subset of the wider marine environment. This article documents the views of fisheries scientists and marine conservationists about innovative approaches towards sustainable fisheries.

TRADITIONAL MANAGEMENT OF FISHERIES

Traditionally, fisheries management and the science underpinning it was distorted by its "narrow focus on target populations and the corresponding failure to account for ecosystem effects leading to declines of species abundance and diversity" and by perceiving the fishing industry as "the sole legitimate user, in effect the owner, of marine living resources." Historically, stock assessment scientists usually worked in government laboratories and considered their work to be providing services to the fishing industry. These scientists dismissed conservation issues and distanced themselves from the scientists and the science that raised the issues. This happened even as commercial fish stocks deteriorated, and even though many governments were signatories to binding conservation agreements.

Defining sustainability

The notion of sustainable development is sometimes regarded as an unattainable, even illogical notion because development inevitably depletes and degrades the environment.

Ray Hilborn, of the University of Washington, distinguishes three ways of defining a sustainable fishery.

- Long term constant yield is the idea that undisturbed nature establishes a steady state that changes little over time. Properly done, fishing at up to maximum sustainable yield allows nature to adjust to a new steady state, without compromising future harvests. However, this view is naive, because constancy is not an attribute of marine ecosystems, which dooms this approach. Stock abundance fluctuates naturally, changing the potential yield over short and long term periods.
- Preserving intergenerational equity acknowledges natural fluctuations and regards as unsustainable only practices which damage the genetic structure destroy habitat, or deplete stock levels to the point where rebuilding requires more than a single generation. Providing rebuilding takes only one generation, overfishing may be economically foolish, but it is not unsustainable. This definition is widely accepted.
- Maintaining a biological, social and economic system considers the health of the human ecosystem as well as the marine ecosystem. A fishery which rotates among multiple species can deplete individual stocks and still be sustainable so long as the ecosystem retains its intrinsic integrity. Such a definition might consider as sustainable fishing practices that lead to the reduction and possible extinction of some species.

Social sustainability

Fisheries and aquaculture are, directly or indirectly, a source of livelihood for over 500 million people, mostly in developing countries. While biodiversity is important, people need food security.

Social sustainability can conflict with biodiversity. A fishery is socially sustainable if the fishery ecosystem maintains the ability to deliver products the society can use. Major species shifts within the ecosystem could be acceptable as long as the flow of such products continues. Humans have been operating such regimes for thousands of years, transforming many ecosystems, depleting or driving to extinction many species. According to Hilborn, the "loss of some species, and indeed transformation of the ecosystem is not incompatible with sustainable harvests." For example, in recent years, barndoor skates have been caught as bycatch in the western Atlantic. Their numbers have severely declined and they will probably go extinct if these catch rates continue. Even if the barndoor skate goes extinct, changing the ecosystem, there could still be sustainable fishing of other commercial species.

Reconciling Fisheries with Conservation

At the Fourth World Fisheries Congress in 2004, Daniel Pauly asked, "How can fisheries science and conservation biology achieve a reconciliation?", then answered his own question, "By accepting each other's essentials: that fishing should remain a viable occupation; and that aquatic ecosystems and their biodiversity are allowed to persist."

Obstacles

Overfishing

Overfishing can be sustainable. According to Hilborn, overfishing can be "a misallocation of societies' resources", but it does not necessarily threaten conservation or sustainability".

Overfishing is traditionally defined as harvesting so many fish that the yield is less than it would be if fishing were reduced. For example, Pacific salmon are usually managed by trying to determine how many spawning salmon, called the "escapement", are needed each generation to produce the maximum harvestable surplus. The optimum escapement is that needed to reach that surplus. If the escapement is half the optimum, then normal fishing looks like overfishing. But this is still sustainable fishing, which could continue indefinitely at its reduced stock numbers and yield. There is a wide range of escapement sizes that present no threat that the stock might collapse or that the stock structure might erode.

On the other hand, overfishing can precede severe stock depletion and fishery collapse. Hilborn points out that continuing to exert fishing pressure while production decreases, stock collapses and the fishery fails, is largely "the product of institutional failure."

A Hubbert linearization (Hubbert curve) has been applied to the whaling industry, as well as charting the price of caviar, which depends on sturgeon stocks. Another example is North Sea cod. Comparing fisheries and mineral extraction tells us that human pressure on the environment is causing a wide range of resources to go through a Hubbert depletion cycle.

Habitat modification

Nearly all the world's continental shelves, and large areas of continental slopes, underwater ridges, and seamounts, have had heavy bottom trawls and dredges repeatedly dragged over their surfaces. For fifty years, governments and organizations, such as the Asian Development Bank, have encouraged the fishing industry to develop trawler fleets. Repeated bottom trawling and dredging literally flattens diversity in the benthic habitat, radically changing the associated communities.

Climate change

Rising ocean temperatures and ocean acidification are radically altering aquatic ecosystems. Climate change is modifying fish distribution and the productivity of marine and freshwater species. This has impacts on the sustainability of fisheries and aquaculture, on the livelihoods of the communities that depend on fisheries, and on the ability of the oceans to capture and store carbon (biological pump). The effect of sea level rise means that coastal fishing communities are in the front line of climate change, while changing rainfall patterns and water use impact on inland (freshwater) fisheries and aquaculture.

Role of Oceans

Oceans and coastal ecosystems play an important role in the global carbon cycle and have removed about 25% of the carbon dioxide emitted by human activities between 2000 and 2007 and about half the anthropogenic CO_2 released since the start of the industrial revolution. Rising ocean temperatures and ocean acidification means that the capacity of the ocean carbon sink will gradually get weaker, giving rise to global concerns expressed in the Monaco and Manado Declarations. Healthy ocean ecosystems are essential for the mitigation of climate change.

Impact on fish Production

The rising ocean acidity makes it more difficult for marine organisms such as shrimps, oysters, or corals to form their shells – a process known as calcification. Many important animals, such as zooplankton, that forms the base of the marine food chain have calcium shells. Thus the entire marine food web is being altered – there are 'cracks in the food chain'. As a result, the distribution, productivity, and species composition of global fish production is changing, generating complex and inter-related impacts on oceans, estuaries, coral reefs, mangroves and sea grass beds that provide habitats and nursery areas for fish. Changing rainfall patterns and water scarcity is impacting on river and lake fisheries and aquaculture production.

Impact on fishing Communities

Coastal and fishing populations and countries dependent on fisheries are particularly vulnerable to climate change. Low-lying countries such as the Maldives and Tuvalu are particularly vulnerable and entire communities may become the first climate refugees. Fishing communities in Bangladesh are subject not only to sea-level rise, but also flooding and increased typhoons. Fishing communities along the Mekong river produce over 1 million tons of basa fish annually and livelihoods and fish production will suffer from saltwater intrusion resulting from rising sea level and dams.

Fisheries and aquaculture contribute significantly to food security and livelihoods. Fish provides essential nutrition for 3 billion people and at least 50% of animal protein and minerals to 400 million people from the poorest countries. Over 500 million people in developing countries depend, directly or indirectly, on fisheries and aquaculture for their livelihoods - aquaculture is the world's fastest growing food production system, growing at 7% annually and fish products are among the most widely traded foods, with more than 37% (by volume) of world production traded internationally.

Adaptation and mitigation

The impacts of climate change can be addressed through adaptation and mitigation. The costs and benefits of adaptation are essentially local or national, while the costs of mitigation are essentially national whereas the benefits are global. Some activities generate both mitigation and adaptation benefits, for example, the restoration of mangrove forests can protect shorelines from erosion and provide breeding grounds for fish while also sequestering carbon.

Adaptation

Several international agencies, including the World Bank and the Food and Agriculture Organization have programs to help countries and communities adapt to global warming, for example by developing policies to improve the resilience of natural resources, through assessments of risk and vulnerability, by increasing awareness of climate change impacts and strengthening key institutions, such as for weather forecasting and early warning systems. The World Development Report 2010 - Development and Climate Change, Chapter 3 shows that reducing overcapacity in fishing fleets and rebuilding fish stocks can both improve resilience to climate change and increase economic returns from marine capture fisheries by US$50 billion per year, while also reducing GHG emissions by fishing fleets. Consequently removal of subsidies on fuel for fishing can have a double benefit by reducing emissions and promoting overfishing.

Investment in sustainable aquaculture can buffer water use in agriculture while producing food and diversifying economic activities. Algal biofuels also show potential as algae can produce 15-300 times more oil per acre than conventional crops, such as rapeseed, soybeans, or jatropha and marine algae do not require scarce freshwater. Programs such as the GEF-funded Coral Reef Targeted Research provide advice on building resilience and conserving coral reef ecosystems, while six Pacific countries recently gave a formal undertaking to protect the reefs in a biodiversity hotspot – the Coral Triangle.

Mitigation

The oceans have removed 50% of the anthropogenic CO_2, so the oceans have absorbed much of the impact of climate change. The famous White Cliffs of Dover illustrate how the ocean captures and buries carbon. These limestone cliffs are formed from the skeletons of marine plankton called coccoliths. Similarly, petroleum formation is attributed largely to marine and aquatic plankton further illustrating the key role of the oceans in carbon sequestration.

Exactly how the oceans capture and bury CO_2 is the subject of intense research by scientists worldwide, such as the Carboocean Project. The current level of GHG emissions means that ocean acidity will continue to increase and aquatic ecosystems will continue to degrade and change. There are feedback mechanisms involved here. For example, warmer waters can absorb less CO_2, so as ocean temperatures rise some dissolved CO_2 will be released back into the atmosphere. Warming also reduces nutrient levels in the mesopelagic zone (about 200 to 1000 m deep). This in turn limits the growth of diatoms in favour of smaller phytoplankton that are poorer biological pumps of carbon. This inhibits the ability of the ocean ecosystems to sequester carbon as the oceans warm. What is clear, is that healthy ocean and coastal ecosystems are necessary to continue the vital role of the ocean carbon sinks, as indicated, for example, by the Blue Carbon assessment prepared by UNEP and the coastal carbon sinks report of IUCN and growing evidence of the role of fish biomass in the transport of carbon from surface waters to the deep ocean.

While the various carbon finance instruments include restoration of forests (REDD) and producing clean energy (emissions trading), few address the need to finance healthy ocean and aquatic ecosystems although these are essential for continued uptake of CO_2 and GHGs. The scientific basis for ocean fertilization – to produce more phytoplankton to increase the uptake of CO_2 – has been challenged, and proposals for burial of CO_2 in the deep ocean have come under criticism from environmentalists. The debate on these issues has underlined the need to increase scientific understanding of how the ocean sequesters carbon.

Ocean pollution

A recent survey of global ocean health concluded that all parts of the ocean have been impacted by human development and that 41 percent has been fouled with human polluted runoff, overfishing, and other abuses. Pollution is not easy to fix, because pollution sources are so dispersed, and are built into the economic systems we depend on.

The United Nations Environment Programme (UNEP) mapped the impacts of stressors such as climate change, pollution, exotic species, and over-exploitation of resources on the oceans. The report shows at least 75 percent of the world's key fishing grounds may be affected.

Diseases and toxins

Large predator fish contain significant amounts of mercury, a neurotoxin which can affect fetal development, memory, mental focus, and produce tremors. Separately, the release of wastewater polluted with microbes, hormones, human and animal medicines and other contaminants can kill fish, reducing sustainable harvests.

Ecosystem based fisheries

According to marine ecologist Chris Frid, the fishing industry points to marine pollution and global warming as the causes of recent, unprecedented declines in fish populations. Frid counters that overfishing has also altered the way the ecosystem works. "Everybody would like to see the rebuilding of fish stocks and this can only be achieved if we understand all of the influences, human and natural, on fish dynamics." He adds: "fish communities can be altered in a number of ways, for example they can decrease if particular-sized individuals of a species are targeted, as this affects predator and prey dynamics. Fishing, however, is not the sole cause of changes to marine life—pollution is another example....No one factor operates in isolation and components of the ecosystem respond differently to each individual factor."

The traditional approach to fisheries science and management has been to focus on a single species. This can be contrasted with the ecosystem-based approach. Ecosystem-based fishery concepts have been implemented in some regions. In a 2007 effort to "stimulate much needed discussion" and "clarify the essential components" of ecosystem-based fisheries science, a group of scientists offered the following ten commandments for ecosystem-based fisheries scientists

- • Keep a perspective that is holistic, risk-adverse and adaptive.
- Maintain an "old growth" structure in fish populations, since big, old and fat female fish have been shown to be the best spawners, but are also susceptible to overfishing.
- Characterize and maintain the natural spatial structure of fish stocks, so that management boundaries match natural boundaries in the sea.
- Monitor and maintain seafloor habitats to make sure fish have food and shelter.
- Maintain resilient ecosystems that are able to withstand occasional shocks.
- Identify and maintain critical food-web connections, including predators and forage species.
- Adapt to ecosystem changes through time, both short-term and on longer cycles of decades or centuries, including global climate change.
- Account for evolutionary changes caused by fishing, which tends to remove large, older fish.
- Include the actions of humans and their social and economic systems in all ecological equations.

Laws and treaties

International laws and treaties related to marine conservation include the 1966 Convention on Fishing and Conservation of Living Resources of the High Seas. United States laws related to marine conservation include the 1972 Marine Mammal Protection Act, as well as the 1972 Marine Protection, Research and Sanctuaries Act which established the National Marine Sanctuaries program. Magnuson-Stevens Fishery Conservation and Management Act.

Awareness campaigns

Various organizations promote sustainable fishing strategies, educate the public and stakeholders, and lobby for conservation law and policy. The list includes the Marine Conservation Biology Institute and Blue Frontier Campaign in the U.S., The U.K.'s Frontier (the Society for Environmental Exploration) and Marine Conservation Society, Australian Marine

Conservation Society, International Council for the Exploration of the Sea (ICES), Langkawi Declaration, Oceana, PROFISH, and the Sea Around Us Project, International Collective in Support of Fishworkers, World Forum of Fish Harvesters and Fish Workers, Frozen at Sea Fillets Association and CEDO.

The United Nations Millennium Development Goals include, as goal #7: target 2, the intention to "reduce biodiversity loss, achieving, by 2010, a significant reduction in the rate of loss", including improving fisheries management to reduce depletion of fish stocks.

Some organizations certify fishing industry players for sustainable or good practices, such as the Marine Stewardship Council and Friend of the Sea.

Other organizations offer advice to members of the public who eat with an eye to sustainability. According to the marine conservation biologist Callum Roberts, four criteria apply when choosing seafood:

- Is the species in trouble in the wild where the animals were caught?
- Does fishing for the species damage ocean habitats?
- Is there a large amount of bycatch taken with the target species?
- Does the fishery have a problem with discards—generally, undersized animals caught and thrown away because their market value is low?

The following organizations have download links for wallet-sized cards, listing good and bad choices:

- Monterey Bay Aquarium Seafood Watch, USA
- Blue Ocean Institute, USA
- Marine Conservation Society, UK
- Australian Marine Conservation Society
- The Southern African Sustainable Seafood Initiative

ENVIRONMENTAL IMPACT OF FISHING

The environmental impact of fishing can be divided into issues that involve the availability of fish to be caught, such as overfishing, sustainable fisheries, and fisheries management; and issues that involve the impact of fishing on other elements of the environment, such as by-catch.

These conservation issues are part of marine conservation, and are addressed in fisheries science programs. There is a growing gap between how many fish are available to be caught and humanity's desire to catch them, a problem that gets worse as the world population grows.

Similar to other environmental issues, there can be conflict between the fishermen who depend on fishing for their livelihoods and fishery scientists who realise that if future fish populations are to be sustainable then some fisheries must reduce or even close.

The journal Science published a four-year study in November 2006, which predicted that, at prevailing trends, the world would run out of wild-caught seafood in 2048. The scientists stated that the decline was a result of overfishing, pollution and other environmental factors that were reducing the population of fisheries at the same time as their ecosystems were being degraded. Yet again the analysis has met criticism as being fundamentally flawed, and many fishery management officials, industry representatives and scientists challenge the findings, although the debate continues. Many countries, such as Tonga, the United States, Australia and New Zealand, and international management bodies have taken steps to appropriately manage marine resources.

Some fishing techniques also may cause habitat destruction. Dynamite fishing and cyanide fishing, which are illegal in many places, harm surrounding habitat. Bottom trawling, the practice of pulling a fishing net along the sea bottom behind trawlers, removes around 5 to 25% of an area's seabed life on a single run. A 2005 report of the UN Millennium Project, commissioned by UN Secretary-General Kofi Annan, recommended the elimination of bottom trawling on the high seas by 2006 to protect seamounts and other ecologically sensitive habitats.

In mid October 2006, U.S. President Bush joined other world leaders calling for a moratorium on deep-sea trawling, a practice shown to often have harmful effects on sea habitat and, hence, on fish populations.

Overfishing has also been widely reported due to increases in the volume of fishing hauls to feed a quickly growing number of consumers. This has led to the breakdown of some sea ecosystems and several fishing industries whose catch has been greatly diminished. The extinction of many species has also been reported. According to an FAO estimate, over 70% of the world's fish species are either fully exploited or depleted. According to Nitin Desai, Secretary General of the 2002 World Summit on Sustainable Development, "Overfishing cannot continue, the depletion of fisheries poses a major threat to the food supply of millions of people."

The cover story of the May 15, 2003 issue of the science journal Nature – with Dr. Ransom A. Myers, an internationally prominent fisheries biologist (Dalhousie University, Halifax, Canada) as the lead author – was devoted to a summary of the scientific information. The story asserted that, as compared with 1950 levels, only a remnant (in some instances, as little as 10%) of all large ocean-fish stocks are left in the seas. These large ocean fish are the species at the top of the food chains (e.g., tuna, cod, among others). However, this article was subsequently criticized as being fundamentally flawed, although much debate still exists (Walters 2003; Hampton et al. 2005; Maunder et al. 2006; Polacheck 2006;Sibert et al. 2006) and the majority of fisheries scientists now consider the results irrelevant with respect to large pelagics (the open seas).

Fishing may disrupt food webs by targeting specific, in-demand species. There might be too much fishing of prey species such as sardines and anchovies, thus reducing the food supply for the predators. It may also cause the increase of prey species when the target fishes are predator species such as salmon and tuna. Fisheries can reduce fish stocks that cetaceans rely on for food.

Many governments and intergovernmental bodies have implemented fisheries management policies designed to curb the environmental impact of fishing. Fishing conservation aims to control the human activities that may completely decrease a fish stock or washout an entire aquatic environment. These laws include the quotas on the total catch of particular species in a fishery, effort quotas (e.g., number of days at sea), the limits on the number of vessels allowed in specific areas, and the imposition of seasonal restrictions on fishing.

In 2008 a large scale study of fisheries that used individual transferable quotas and ones that didn't provided strong evidence that individual transferable quotas can help to prevent collapses and restore fisheries that appear to be in decline.

Fish farming has been proposed as a more sustainable alternative to traditional capture of wild fish. However, fish farming has been found to have negative impacts on nearby wild fish. Further, farming of predatory fish like salmon can rely on fish feed that is based on fish meal and oil from wild fish. The environmental impact of recreational fishing may be alleviated to some extent by catch and release fishing.

FISHING DOWN THE FOOD WEB

The idea that we on a global level were Fishing down the food web was first proposed by the fisheries scientist Daniel Pauly and others in an article published in the journal Science in 1998. Large predator fish with higher trophic levels have been depleted in wild fisheries. As a result, the fishing industry has been "fishing down the food web", targeting fish species with trophic levels that are progressively decreasing. The authors were criticized for the mismatch between the simplicity of the modelling approach and the categorical conclusions by Caddy et al. in a later Science issue the same year. A more elaborated review contradicting Pauly et al. appeared in 2010, indicating that the Mean Trophical Level (MTL) in the worlds oceans are stable or increasing Over the last 50 years, the abundance of large predator fish, such as cod, swordfish and tuna, has dropped 90 percent. Fishing vessels now increasingly pursue the smaller forage fish, such as herrings, sardines, menhaden and anchovies, that are lower on the food chain. "We are eating bait and moving on to jellyfish and plankton" says Pauly. Beyond this, the overall global volume of fish captured has been declining since the late 1980s.

Mean trophic level

The mean trophic level is calculated by assigning each fish or invertebrate species a number based on its trophic level. The trophic level is a measure of the position of an organism in a food web, starting at level 1 with primary producers, such as phytoplankton and seaweed, then moving through the primary consumers at level 2 that eat the primary producers to the secondary consumers at level 3 that eat the primary consumers, and so on. In marine environments, trophic levels range from two to five for the apex predators. The mean trophic level can then be calculated for fishery catches by averaging trophic levels for the overall catch using the datasets for commercial fish landings.

Ecopath

Pauly's team used the catch data from the FAO which it fed into an Ecopath model. Ecopath is a computerised ecosystem modelling system. The functioning of an ecosystem can be described using path analysis to track the direction and influence of the many factors controlling the ecosystem. The original Ecopath model was applied to a coral reef food web. Scientists tracked tiger sharks at the top of the food web and collected data on their feeding behaviour, what they ate and how much. Likewise, they collected feeding data on the other organisms in the food chains down to the primary producers, such as algae. This data was fed into an Ecopath model, which then described the energy flow, in terms of food, as it moved from the primary producers up the food web to the apex predator. Such models allow scientists to compute the complex effects that occur, both direct and indirect, from the interactions of the many ecosystem components.

The model showed that over the last 50 years the mean trophic level of fish catches has declined by somewhere between 0.5 and 1.0 trophic levels. This decline applied both globally, on a world wide scale, and more locally on a scale specific to oceans, that is, for the separate FAO subareas: the Atlantic, Indian and Pacific Oceans, and the Mediterranean-Black Seas.

Marine Trophic Index

In 2000, the Convention on Biological Diversity, an international treaty aimed at sustaining biodiversity which has been adopted by 193 member countries, selected the mean trophic level of fisheries catch as one of eight indicators for immediate testing. They renamed it the "Marine Trophic Index" (MTI), and have mandated that member countries report over time on changes in ocean trophic levels as a primary indicator of marine biodiversity and health.

The "Marine Trophic Index" is a measure of the overall health and stability of a marine ecosystem or area. The index is also a proxy measure for overfishing and an indication of how abundant and rich the large, high trophic level fish are.

Changes in the Marine Trophic Index over time can function as an indicator of the sustainability of a country's fish resources. It can indicate the extent that the fishing effort within a country's fishing grounds is modifying its fish stocks. A negative change generally indicates that larger predator fish are becoming depleted, and an increasing number of smaller forage fish are being caught. A zero or positive change in the Marine Trophic Index indicates the fishery is stable or improving.

FiB Index

The ecology of fishing down

Ecologically, the decline in the mean trophic level is explained by the relationship between the size of the fish captured and their trophic level. The trophic level of fishes usually increases with their size, and fishing tends to selectively capture the larger fishes. This applies both between species as well as within species. When the fishing is intense, the relative abundance of the larger fish positioned high in the food chain is reduced. Consequently, over time, small fishes start to dominate the fisheries catches, and the mean trophic level of the catches declines. Recently the market value of small forage fishes and invertebrates, which have low trophic levels, has sharply increased to the point where they can be considered to be subsidizing fishing down.

Daniel Pauly has suggested a framework for the ecological impacts fishing down can have on marine ecosystems. The framework distinguishes three phases:

- Pristine – the first phase. Pristine environments are the states ocean ecosystem were in before fishing made strong impacts. Some outlier areas of the South Pacific may still be pristine. For most of the world what these pristine states may have been can only be inferred from archeological data, historical accounts and anecdotes. In pristine ocean environments, the biomass of the large predator fish are 10 to 100 times greater than their present biomass. This implies a large supporting biomass of small prey fishes and invertebrates. On the seafloor, the benthos is dominated by deposit feeders which prevent resuspension of sediments, and filter feeders which keep the phytoplankton down. Thus the water column tends to be oligotrophic, free of both suspended particles and of the nutrients that leach from them.

- Exploited – the second phase. Exploited is the phase we are currently in. It is characterized by declines in the biomass of large predator fish, declines in the diversity, size and trophic level of captured fish, and declines in the benthos. Bottom trawlers progressively destroy the biogenic structures built over many years on the seafloor by the filter and detritus feeders. As these structures and animals that were filtering the phytoplankton and consuming the detritus (marine snow) disappear, they are replaced with the polyp stages of jellyfish and other small errant benthic animals. Storms resuspend the marine snow, and the water column gradually eutrophies. In the early part of this phase, cascade effects compensate these declines with the emergence of new fisheries for opportunistic feeders, such as squid, shrimp and other invertebrates. But eventually these decline also.
- Fully degraded – the third phase. The dead zone is the biological endpoint of a fully degraded marine ecosystem. The dead zone is a zone with excessive nutrients in the water column, resulting in the depletion of oxygen and the elimination of multicellular organisms. The abundant detritus and marine snow is processed by bacteria rather than by the benthic animals. These dead zones are currently growing throughout the world in places such as the Bohai Sea in China, the northern Adriatic Sea, and the northern Gulf of Mexico. Some estuaries, such as the Chesapeake Bay estuary, also display features associated with a fully degraded marine ecosystem. In Chesapeake Bay, overfishing eliminated the benthic filter feeders such as oysters, and most predators larger than a striped bass, the current apex predator. One hundred and fifty years ago, the oysters formed giant reefs and filtered the waters of Chesapeake Bay every three days. Because the oysters are gone, pollution entering the estuary from rivers now produces harmful algal blooms.

Fishing through the web

- Essington† TE, Beaudreau AH, Wiedenmann J (2006) "Fishing through marine food webs" PNAS, 103 (9): 3171-3175.

- Shopping list gets longer — not less choosy — in some of world's largest fisheries University of Washington news, February 14, 2006.
- National Research Council (Authors) (2006) Dynamic Changes in Marine Ecosystems: Fishing, Food Webs, and Future Options National Academies Press. ISBN 978-0-309-10050-2

Farming up the web

While the mean tropic level in wild fisheries has been decreasing, the mean tropic level amongst farmed fish has been increasing.

As an example, the table above shows trends in the trophic levels of fish farmed in the Mediterranean. However, the farming of bluefin tuna is restricted to a fattening process. Juvenile tuna are captured from the wild and put in pens for fattening. Wild stocks of bluefin are now threatened, and the fisheries scientist Konstantios Stergiou and colleagues argue that the "fact that the capacity of tuna farms greatly exceeds the total allowable catch indicates lack of conservation planning in development of the tuna-fattening industry, which, ideally, should have been linked to fisheries management policies, and may lead to illegal fishing."

Also, fish farming in the Mediterranean is a net fish consumer. Large amounts of animal feed is needed to feed a high trophic fish like the bluefin tuna. This feed consists of fishmeal processed from forage fishes like sardines and anchovies that humans would otherwise consume directly. In addition to ecological issues, this raises ethical issues. Much of the fish suitable for direct human consumption is being used to grow higher trophic level fish to indulge a relatively small group of affluent consumers.

DESTRUCTIVE FISHING PRACTICES

The phrase destructive fishing practices (or DFPs) has been featured in international fisheries literature for around three decades. No widely accepted definition of the phrase exists, and this will almost certainly remain the situation, given very different national and industry perspectives. The DFPs discussion with the most standing is that of the Fisheries and Agriculture Organization of the United Nations - the FAO.

The Outcomes and Implementation Statements of the World Summit on Sustainable Development, held in Johannesburg in 2002, contain a commitment to phasing out destructive fishing practices in the marine environment by the year 2012. All nations attending the summit supported this statement.

Many nations had made commitments to end destructive fishing practices much earlier. In 1999, 124 nations explicitly gave their support to the FAO Code of Conduct for Responsible Fisheries 1995 through the Rome

Declaration on Responsible Fisheries. The list of these nations includes most of the major fishing nations of the world. However, while the Code of Conduct contains a commitment to end destructive fishing practices, the Code contains no time-lines.

The narrowest definition of destructive fishing practices refers principally to bottom trawling over vulnerable habitat (shallow corals, deep sea corals , or seagrass, for example), as well as practices such as shark-finning, blast-fishing, poison-fishing, muroami , and push-netting. These latter practices are not significant within the fishing zones of most developed nations, being generally outlawed. Bottom trawling over vulnerable habitat, however, will continue within the Exclusive Fishing Zones of most nations until governments have mapped the location of vulnerable habitats, and taken steps to exclude all bottom trawling activities from these areas.

A wider and more useful definition would include:

- overfishing beyond reasonable recovery limits , including serial overfishing;
- excessive or damaging levels of bycatch ;
- The fishing of spawning aggregations without precautionary justification;
- Intensive fishing over vulnerable habitats, including for example spawning and nursery areas; and
- highgrading-the practice of discarding the lower quality portion of the target catch.

This definition could be extended to cover activities such as:

- ghost fishing by lost or discarded gear ,
- shark netting of popular swimming beaches (with high incidental catch),
- Amateur use of fish aggregating devices or traps where they increase the likelihood of locally unsustainable catch levels,
- spearfishing at night or with SCUBA, as well as 'industrial' spearfishing,
- use of stainless steel hooks or traps;
- commercial and recreational use of gill nets and traps with high incidental kill, and
- deliberate (and sometimes illegal) destruction of marine life perceived as "getting in the way" of fishing operations .

2

FRESH WATER FISH CULTURE

The term fresh water refers to water with neutral quality (PH around 7), clear, clean, colorless, well oxygenated, having normal temperature (ranging between 20 to 30 degree centigrade). It may be flowing or not flowing as in the case of wells, ponds, lakes reservoirs etc. The nonflowing water is also not really a non flowing water. In ponds, wells, lakes and reservoirs also there is always the movement of water due to seepage, percolation, under ground water flow at the level of water table, evaporation and precipitation through rain, hail stones or even by snowing.

This movement may be both incoming and outgoing. These movements of water arc not seen and therefore we say that the water in these water bodies is nonflowing. In fact it is these types of incoming and outgoing movements, that maintain the water fresh in these bodies. The fresh water in the tropical regions may be referred to as warm water as compared to cold water in the temperate regions. Otherwise the real meaning of warm water refers to water above 30 to 35 degree centigrade. Above this temperature normally the cultivated fishes do not thrive well. Rain water is the best form of fresh water in the nature.

Based on the temperature tresh water can be of two types: water with normal temperature and cold water. Climate wise it is same as waters in the tropical and temperate areas. There are a number of ways of culturing fishes and other aquatic organisms and plants in the fresh water with normal temperature and below normal temperature or cold though most of the fishes can be reared both in the temperate and tropical regions. They are briefly introduced here.

1. Carp culture

Rearing fishes belonging to the carp or Cyprinid family is carp culture. There are about 25 species in the carp family that are recommended for rearing allover the world. Most famous among them are the Indian, Chinese and European carps for their quick growth and suitability for composite fish farming. Carla, rohu, mrigal have the Indian carps while the Chinese counter parts are silver carp, grass carp and common carp. Carp (Cyprinus caprio), mirror carp, leather carp, tench, roach, rudd, crucial carp, gold fish and golden ide are some of the European carps which are mostly grown in the temperate climate.

The Chinese and Indian carps are reared both in the tropical and temperate climate. They mature in two years period and are easy to breed artificially in the ponds by the injection of the pituitary gland They also respond well to supplementary feeding. Catla and silver carp feed on the upper layers of the water, rohu and grass carp do feed at the middle layers and mrigal and common carp teed from the bottom layers of the ponds. Because of these characteristics these fishes fit in very well with each other in a composite fish culture.

2. *Tilapia culture*

Tilapia, belonging to Cichlidae is a versatile fish that can be reared in any fresh water condition. They are omnivorous and hence feeding is least problematic. They breed and multiply by themselves and hence raising fry and fingerlings need not be a concern for the farmer. They multiply so fast that in no time the pond may be full of small fish and there will be competition for the feed and space. Besides, the growth rate will be drastically reduced. To solve this problem farmers are advised to keep carnivorous fish like catfish which will eat up quite a large number of the fry and keep the population of the tilapia under control.

· Another method employed to prevent the undue multiplication of the tilapia is to breed in such a way that the result- ant progeny will be only the male tilapia. Rearing only the male tilapia we can avoid the problems of uncontrolled breeding and multiplication.

Tilapia in combination with the cat fish is very suitable for those people who have no breeding facilities, no facilities to procure fry and fingerlings, no supplementary feed availability and poor marketing facilities.

3. *Catfish culture*

Cat fish, belonging to Calaridae family is an air breathing fish and hence they can survive in areas where there is less water depth or even less oxygenated water. They are commonly found in all the waterlogged areas of the tropical climate. Compared to most other types of fishes, they need very little care. Though carnivorous by nature they feed on all kinds of dead and living plants and animal material and hence feeding is easier. As already mentioned catfish can -be reared along with tilapia in order to control the undue breeding and multiplication of tilapia.

The system of catfish culture is recommended in the following situations.

a. Marshy and water logged areas where coconut is cultivated on big broad bunds or strips of land with intermittent long ditches. The width of the bunds and ditches may range between 10 to 30 ft and the length may be adjusted as per convenience or availability of the land. The width of the bund is more in areas where the quantity of water is less; whereas if the water is more the width of the ditch or channel will be more than the bunds on which the coconut .stands.

b. Where the water table is very high catfish can be reared in narrow ditches or channels intermittent with strips or blocks of banana plantation.

c. In paddy fields and adjacent canals the catfish can be easily reared.

d. In channels or ditches between sugarcane fields the cat-fish can be reared.

e. In natural ponds and small lakes which are overgrown with aquatic plants the catfish grows naturally.

f. In sewage collection ponds or tanks catfish, being an air breathing fish can be grown.

In short wherever natural ditches and ponds with fresh or less fresh water are there, catfish can be grown easily.

4. *Salmon culture*

Salmon belonging to the farnily Ofsalmonidae is cultivated in cold water throughout the world. But they are usually reared in the western countries and in regions where the temperate climate is prevalent. Though there are many species of salmon in the world, about twenty are recommended for rearing in the ponds. In the west the system of salmon culture is highly developed and common. In India salmon can be cultivated in regions where temperate climate is prevalent.

Salmons are mainly of two types: one is called salmon and the other called trouts though the family is the same. Hence we can consider the culture of salmon and trouts separately though many of the rearing practices are the same.

5. *Trout culture*

Trout is a cold water fish and is much valuable compared to other fishes. They inhabit mainly the rivers, big and small lakes and ponds of upland areas like Kashmir, Himachal Pradesh, Uttar Pradesh, Nilgiris, Kodai hills and high ranges of Kerala. They have a streamlined body, narrow gill openings and reduced gills and are adapted to highly oxygenated water and freezing point temperature. Trouts can be reproduced under artificial conditions like the carps for the production of fry and fingerlings.

Comparatively trouts require fresh and clean water with high oxygen content. The dissolved oxygen under lower temperature is higher, when compared to oxygen content in the water under the higher temperature. In the hills the temperature is lower and hence the oxygen content in the water in those areas is more.

Trouts are easy to manage and harvest. Their production capacity is 100 to 200 kg/ha in the pond conditions. The same productivity can be maintained in the waterways, canals, small streams, and other small water bodies etc.

Trouts in general especially the rainbow trouts are exceptionally tolerant to overcrowding. Hence they can be reared even in drums and other big containers provided a good flow of well oxygenated fresh water at the rate of about 12 litres per minute is maintained. Drums that receive 25 litres of fresh water per minute utilizes the water for self cleaning also. However once a month cleaning of the faecal matter and other debris is important.

6. Eel Culture

Eels are snake like fishes but very valuable in the world market. They are nocturnal in their habit and are generally carnivores. Out of the 19 known species, 17 are found in the Pacific and Indian ocean. Indian ocean alone has about six species. They migrate into the rivers to grow and mature to go back again into the sea to breed so that another generation of eels would migrate again.

Experiments in Japan and other countries have shown that they can be reared and bred in captivity. This offers tremendous opportunity for large scale eel culture in our country. However much work need to be done in the breeding and rearing of eels in our country.

7. Culture of other fishes

Similar to carp, tilapia, catfish and salmon we can have systems of rearing of individual fish breeds such as milk fish, shad, sturgeon, black bass etc. Thus there can be as many systems of fish production as the breeds of fishes are.

8. Fish culture in cages

Fish can be reared in cages and enclosures immersed in pond, stream, river lakes, well or any other water body. This has been in practice for many years in some of the South Asian countries where there is plenty of interlinked water bodies. The fish is kept in the cage and the cage is kept immersed in water. By this method one can rear fish even in common water bodies such as rivers, streams, lakes, canals and ponds. Thus people who do not have any land for ponds can also rear fish.

The enclosures are built in any of the above mentioned water bodies at a suitable place in convenient shape or size. In the water bodies there will be an inlet and an outlet to maintain the minimum flow of water. The outlets and inlets are fixed with wire mesh so that the fish inside will not escape.

The species traditionally reared in the cages are common carp, catfish, salmon, trout, channel catfish and milk fish. However the air breathing fishes are more suited to cage rearing.

9. Paddy cum fish culture

Paddy is mostly a fresh water crop that grow well in flooded conditions. with certain modifications in the structure of the paddy field we can easily rear some selected fish varieties in the paddy field. In permanently flooded paddy fields fish culture can be done for one or two years continuously before one can harvest them. But seasonally flooded paddy fields provide comer ponds for the fish to remain safe during the dry season or during the harvesting time. The paddy field should be surrounded by strong bund high enough or some sort of closely made fencing to prevent the fish from escaping from the field.

When rearing fish in the paddy field we cannot use any kind of pesticides or weedicide as they may kill the fish or harm them. The structuring of the paddy field and the rearing methods followed depends on the ingenuity of the fish cum paddy farmer.

10. Fish culture with animal husbandry

Animal dung, urine and animal shed wash are excellent organic manures to fertilize the fish ponds. They promote the micro flora and fauna which are the natural feed for most of the fishes. Most of these are not visible to the naked eye and hence people do not value them as they should. Besides the micro organisms these organic manures encourage higher level flora (various kind of water plants) and fauna (various kinds of aquatic insects and worms). For some fishes the decaying organic matter itself is a feed.

Above all due to the application of organic manures in the pond aquatic system becomes alive and active and behaves like an organic unit. Fishes in such an aquatic ecosystem will thrive better. Thus there exists a beautiful symbiotic relationship between the animal units and the fish ponds in it farm. 'This relationship can he exploited for better economic returns at various levels at the home level and the commercial and scientific level.

Under such situations one has to foresee that the animal units and the fish ponds are constructed in such a way that this symbiotic relationship can easily be made operative in the farm. It is achieved by constructing them sufficiently close to each other so that dung, urine and animal shed wash can be easily transferred to the fish ponds whenever one needs. At the same time all the dung, urine and shed wash are not sent to the ponds. Such a practice will result in over fertilization of the ponds creating unhealthy conditions to the fishes. Obviously the ponds are fertilized only according to the requirements at a particular time.

In a fish farm there are lots of broad bunds and vacant spaces in between the ponds and all around the pond areas for growing fodder grasses and trees which are additional resources for incorporating animal husbandry units. Depending on various factors (availability of foundation stock, feed, veterinary help, marketing etc.) that contribute to the success of the animal units one can start any or all of the animal units.

Cattle, goat, sheep, pigs, chicken, ducks, geese, guinea fowls; turkeys, ostriches, rabbits, guinea pigs, horses, donkeys are the most common animals we can rear in the fish farms. The sheds for these animals and the fish ponds are constructed in such a way that the dung, urine and shed wash can be easily transferred to each fish pond separately whenever one wants them in the required quantities. Between each animal unit and the fish pond, there should be a holding place for the dung (compost boxes: Booklet No.146), for the urine and shed wash (urine and shed wash collection tank). Dung is applied into the pond only after composting well whereas the urine mixed with shed wash should be used within three days. Otherwise most of the nitrogen may be lost into the air through ammonia formation.

For the designs of various animal housing units and details about various management practices for various animals, the reader is requested to refer Booklets on various topics covered under various animals. Similarly for details about of fish culture refer Booklets on various topics on the same, because the designs and the layout of both animal housing and fish ponds cannot be simply replicated in all the cases and at all the places. Depending on the location, climate, soil, need, facilities, size and purpose of the animal cum fish culture designs for each unit has to be planned. The compost boxes and the urine and shed wash collection tanks are the intermediary structures between the animal units and the fish ponds.

11. Cold water fish culture

The fishes that thrive in the cold climate are different from those found in the tropical climate. As already mentioned salmon is reared in the cold waters of temperate region of Europe and other countries with temperate climate. In India Kashmir, Himalayan regions and hilly regions like Nigiris are suitable for culture of cold water fishes. The most important species of cold water fishes are those belongs to the salmon family known as tout in general. Among them brown tout, rainbow tout, eastern brook tout, land locked salmon, golden strain and kokanee salmon are commonly cultured. Some of these are introduced into the above mentioned temperate regions of India.

In India touts are cultured in dug-out, running water ponds. The rate and amount of water flow depends mainly on the number and size of the tout stocked and the volume of the water in the pond. Greater the volume of water more should be the water that is to be recycled. The walls of the ponds are reinforced with stone or cement. However tout culture in India is not so successful as the tropical fishes.

12. Culture of ornamental fishes

Rearing of ornamental or aquarium fishes is popular all over the world. Even in individual families people try to rear few ornamental fishes in a small glass box. There are huge aquariums in which a large number of the ornamental fishes are reared for providing entertainment and knowledge. Hence a system of rearing ornamental fishes is itself developed by the scientists.

Based on the type of fish and other organisms being grown the aquaria can be classified into as many types of fish, aquatic animals or organisms as possible. In most cases at the family level, a mixture of different types of ornamental fishes are reared. In the public aquaria different ornamental fishes are reared in different glass cases or cages.

There are a number of aquatic organisms which do not come under fishes but are reared or harvested extensively for food and market purpose. They are mostly crustaceans such as shrimps, prawns, crayfish, crabs, clams etc. These are also found in nonfreezing water. However for logical classification and sequence in this section we shall refer only to those that are found in the fresh water. Fresh water shrimps, prawns, crayfish, crabs and clams are the commonly grown nonfishes. Among them the giant fresh water prawns are economically more important.

III. Non Fresh Water Fish Culture

The non-fresh water is a term used in a very broad sense and refers to all types of water that do not have the characteristics of the fresh water as described already. It may be saline, alkaline, brakish water, sea water, acidic, poorly oxygenated, muddy, sewage water, drainage from the residential areas or from industrial areas, or water in the marshy and swampy areas. It may be found in lagoons, esteries, brakish water lakes, canals, mangroves or inland ponds or wells with saline or brakish water.

There are people who have such types of water under their ownership or control. Several types of fishes can be reared in these types of waters and very good economic return can be obtained by them provided they have the know how. Fish rearing in such types of water may be termed as Non Fresh water Fish Rearing. Some of the most important types of non-fresh water fish culture are briefly introduced here.

1. *Air breathing fish culture*

From the name itself we can understand the nature of the fishes being reared in this system. These fishes can live by breathing the air directly. That means the water in which they live is unsuitable or is unable to provide enough oxygen to these fish because the water may not be really fresh. Such situation occurs in the nature due to many reasons as ex- plained here.

Many low-lying areas where rainfall is fairly heavy and topography of the land is such that water logging occurs, get infested with various types of aquatic vegetation and become swamp. At times large waterlogged areas become choked with weeds which get decayed and the whole area becomes a swamp. In this swamp considerable quantities and qualities of organisms both of the plant and animal types grow and multiply. Various types of micro and macro-phyto and zooplankton (flora and fauna) and decaying organic matter are common in these swamps. Often the water of these swamps are not really fresh and very poor in the oxygen content.

The best way to use these types of swamps is to rear air breathing fishes such as giant murrel (Channa marulius), stripped murrel (C. striatus), spotted murrel (C.punctatus), walking fish or catfish (Clarius batrachus) singhi (Heteropneustes tassilis) and climbing perch (Anaba testudineus). All these are carnivorous as well as feeders of dead and decaying materials of plant and animal, origin. They also easily breed and multiply in these swampy water bodies.

Monoculture as well as polyculture of these air breathing fishes are done in many of the world where such natural lakes, water logged areas and swamps occur commonly. Efforts are being made under an All India Coordinated Research Project to develop suitable techniques for air-breathing fish culture which have given encouraging results.

There may be enough natural feed generated in these water bodies under the natural conditions. However if possible supplementary feeding, should be given to augment the growth of the fishes.

2. *Sewage water fish culture*

In the present urbanized living conditions we generate large quantities of sewage water from the residential and industrial areas. Most of this can be utilized for agricultural and fishery purpose after proper treatment and dilution. The sewage water is treated and recycled in many parts of the world for a number of purposes. The major steps involved in the treatment of the sewage are the following.

a. Running down the sewage through a number of ponds,

b. Filtration of sewage in two stages followed by sedimentation,

c. Aeration of sewage and

d. Dilution of sewage with fresh water (one part sewage with five parts of fresh water).

The sewage is made to stand in the ponds called oxidation ponds for about 10 days. During this time the heavier particles settle down in the form of sediment. Meantime due the action of aerobic microorganisms, there is break down of the organic matter resulting in the formation of inorganic nutrients. The aerobic decomposition improves the nutrition quality of the sewage water and algae and other microphytes grow and the water becomes suitable for fish culture. By growing water hyacinth, Eicchornia and other water plants in the oxidation ponds we can remove the toxic heavy metals from the water.

The total capacity of the oxidation ponds should be such that they should able to retain at least 1.0 to 12 days sewage. The ponds should not be more than 3-4 ft deep for the light to penetrate and the air to get mixed with for the oxidation process to happen.

For the full mutilation of the treated sewage water the total capacity of the fish ponds should be equal to four to five time that of the sewage. The treated sewage water can tie considered equal to the total sewage let in for treatment minus the loss of water through evoporaton, seepage and transpiration in the case of growing water plants.

Fishes like silver carp, rohu, mrigal, common carp, grass carp, catla, fringe-lipped carp etc. can be reared in the sewage fed fish ponds.

3. *Brackish water fish culture*

Slightly salty water is called brackish water. Such type of water is found in areas where the sea water get mixed with fresh water coming from the rivers and streams joining the sea (esturies) or fresh lake water get mixed up with sea water at the time of high tide. Brackish water is commonly observed in the water bodies situated along the sea shores. Certain fishes and prawns thrive well in the brackish water. Brakish water prawn culture is becoming very popular in India because prawns fetch very high prices.

There are several other fishes that can be reared in the brackish water. But they are not becoming popular and commercial because their economic profits from them are not sufficiently attractive.

Besides various kinds of fishes and prawns several other economically beneficial organisms of both plant origin and animal orgin can be cultured in brackish water. But except for the prawns, markets have not been developed significantly for the brackish water fishes. The scanty knowledge system and technology about rearing water organisms in the brackish water is also another reason for the poor development of the brackish water fish culture.

4. *Marine fish culture*

When we hear about marine fish culture we may wonder how fishes are reared in the sea? Of course huge enclosed areas in the sea can be used for marine fish culture. All that required is to have a constant linkage and mixing of water within the enclosure and the water outside. Such systems of marine fish culture is still only in the conceptual level though Japanese are trying to do something in this line.

The Japanese breed and rear fishlings up to certain age and then they are let into the sea. By doing this they are ensuring survival of the maximum fishlings in the sea so that the harvestable population of fish in the sea is maintained optimum. Otherwise there is always very high mortality of fishlings in the nature and the fishes that grow up to harvestable age is very much reduced.

Normally certain types of fishes are found to thrive well in certain part of the sea. In other words diff'erent types of fishes are seen to be thriving well in different parts of the sea or different sea. For example fishes commonly found in the West coast area are different from the East coast area. Again fishes that are found in the Orissa coast are different from that of in the Tamil Nadu coast. In the same way the fishes found in the Gujarat coast are different from that are found in the Karnataka or Kerala coast. Even in the same state the fishes in one coastal area will be different from the fishes of some other coastal area. For example the fishes in the southern parts of Kerala coast are different from those found in the Northern parts of Kerala; similarly the fishes in the sea of Cape Comerin area are different from those found in the Tuticorin or Chennai area. Even in different parts of the marine areas of the same state we find different types of fish. For example the fishes in the marine areas of the Malabar area is different from the fishes of the Cochin or Trivandrum areas.

Taking advantage of this phenomenon in the sea we can increase the fish production in the sea by several ways. The first method is to increase the survival rate of the frys and fingerlings of the fishes. This is done by breeding selected fishes to produce trys (md rearing the frys till they become fingerlings and then releasing them into the sea. By this the survival rate of the fishes in the sea is increased. The second method is by increasing the generation of the natural feed of marine flora and fauna or by supplying supplementary feed for the fishes of a particular area. Fishes will remain in those areas where abundant feed is available. By generating natural feed in a particular area or by providing supplementary feed or by both we can retain the fishes of an area to that place and harvest them when they are sufficiently grown.

Obviously the marine cultivation of fishes is very expensive and probably only the government and the multinationals will be able to carry out such programmes. However it is good for the ordinary people to know that such possibilities are therein marine fish culture.

IV. Non-Fish Aquaculture

Besides the above mentioned fish cultural practices for fishes the aquacultural practices include culture of several organisms other than the fishes. Though some of them are already being practiced commercially they are yet to be widely accepted by the people as possible and profitable aquaculture practices. Already we have some knowledge about most of them so that anyone can start and experiment on any of them. This portion of the booklet compiles whatever knowledge we have on some of the possible non-fish aquacultural practices such as shrimps, prawns, crayfish, lobsters, clam, frog, crab, crocodile, tortoise, pearl culture, etc. In this section no distinction is made between the fresh water and non fresh water aquaculture as in the case of fish rearing.

1. Shrimps, prawns, cray fish and lobesters

In the common parlance people seem to confuse between shrimps, prawns, crayfish and lobsters because externally they look much alike to an ordinary person. Actually they are different physically and genetically. To distinguish the differences between them one has to observe carefully the anatomical differences, Size wise shrimps are small while the prawns and the crayfish are bigger and the lobsters are the biggest. Shrimps have thin and transparent shells/coverings while the prawns, crayfish and lobsters have thick and colored coverings/shells, There is a lot of variation in the colours and colour patterns among the prawns, crayfish and lobesters. The legs and antennae of the shrimps are very slender and tender while that of the prawns and crayfish are harder and stronger and that of the lobesters hardest. The body of the shrimps is very smooth and that of the lobtsters most rough and spiny while the prawns and the crayfish range in between. The shrimps do not appear to have any pincers while the prawns and crayfish have their first pair of legs turned into pincers. The pincers of the crayfish are broader and shorter (like that of a scorpion) than that of the prawns which have long and narrow pincers. Whereas in the case of lobesters more than one legs are turned into pincers the first one being very prominent and terrifying.

Habitat-wise the shrimps and prawns grow in fresh and non fresh waters, while the crayfish is only found in fresh water streams and rivers with plenty of sand, gravel and rocks and lobesters are found in the deep sea.

After having described briefly the basic distinguishing characteristics between the shrimps, prawns, crayfish and lobesters we can now focus on their fresh water cultural aspects. However the cultural aspects of the shrimps and prawns are more developed than all others. A lot more work needed be done to generate sufficient theoretical and practical information regarding their life cycle, habitat, growth and behaviour before we can start their commercial rearing. Such information is more available at present on the case of shrimps and prawns.

Though there are many species of shrimps and prawns mainly two species are used for commercial culture: Penaeus monodon and Penaeus indicts. Both take about 15 to 20 days to go through the early stages (eggs, six nauplius stages, two protozoan stages, three mysis stages and five post larval stages and after that by 90 to 100 days to go through early juvenile, sub-adult and adult-immature stages. After this they are in the adult stage and reaches marketable size by 10 to 12 months age. Among the prawns culture of fresh water, giant prawns are more important commercially.

The culture of cray fish and lobesters are still to be taken up in a commercial manner. Perhaps much research is needed to generate sufficient practical know-how regarding the culture of these two. However one can try to select some good breeding stocks and start rearing them.

The success depends on how good a pond or growing situation one has, how far one can ensure maximum survival of the young ones till they become adults and how well one can feed them with nutritious and balanced feed. For the first purpose, one should have a well planned pond or growing condition, for the second purpose, one should have the support of a very good hatchery either own or by some one reliable and for the third one should have a steady supply of proper feed. Shrimps and prawn culture is a highly complicated practice that any one who plans to go for it should have a thorough knowledge on the same. The shrimps and prawns can be cultured both in fresh and non-fresh water.

2. *Mollucs, clam and oyster culture*

Molluscs, clams and oysters are name. applied to several species of bivalves. They can be cultured in their natural envious or under suitable artificial conditions. Clams among them are the most common and are two types: one having the soft and the other having hard shells. The soft shelled belong to the species Maya areneria and the hard shelled to Venus mercenaria. They burrow into the sand or mud. The soft shelled clam is characterized by a stretchable neck or siphon having two tubes: one for sucking in water with food materials and the other for expelling wastes.

The hard shelled clam is strongly hinged, close fitting, connected by two powerful abductor muscles and stays in the mud or sand sheltered by plants or something. Some of the hard shelled clam grow up to a kilogram weight and may live up to 35 years.

Both these type of clams are used for meat and for shell which is good source of calcium. The artificial culturing of clams is yet to be taken up in the commercial level though there is plenty of scope for the same. At present they are mostly harvested from the nature.

The female clams release millions of eggs of which very few get fertilized and of the hatched one (called veliger) only very few survive to adulthood in the natural conditions. Under artificial rearing conditions we can increase the survival rate and production tremendously. They can be cultured both in fresh and nonfresh water conditions.

3. *Frog rearing*

Frog meat is excellent from the point of view of digestibility and nutrition. It is highly valued in the foreign countries. In India people have so much cultural and religious inhibitions for eating frog meat. If they can get over this it will not only help them to overcome the nutritional deficiencies but also enhance their income. Frogs can be reared in simple and small scale ways at home or large and commercial level provided one knows at least the basic requirements of frog culture. However for large scale commercial scale rearing a thorough knowledge about the breeds, breeding and reproduction, life cycle, feeding, habita., rearing methods, pests and predators etc. is essential.

Frog rearing has several advantages: (1) it involves little capital investment (2) frog waste could be processed into a protein rich byproduct similar to fish protein concentrate and used as a nutritional supplement in fish and livestock feeds (3) frog skins could be tanned for the manufacture of leather goods and (4) adult frogs and their larval stages could effectively control disease-spreading insect. Thus frog rearing has very high economic potential which is still to be harnessed.

Three indigenous species: Rana tigrina(Indian bullfrog), R dactyla (Grecn frog) and R. Crassa (Jerdon's frog) are found to be very suitahle for commercial rearing in our country. An exotic frog called R. catesbiana (North America bullfrog) was introduced into our country in 1968 by the Central Fishries Research Institute (CFRI), Cuttack. These four are being cultured in our country at present. However commercial culture is yet to become common. For more details about the frog culture contact (CFRI or refer Booklet No.561 on Frog Culture.

4. *Crab culture*

Crabs are delicacies and very costly. Hence culturing them for home use or for commercial purpose is certainly beneficial for the family nutrition as well as for the high profitability. However crab culture is still to be developed in its theory and practice.

One can find crabs almost at any place where there is water: in the sea, in the lakes, in rivers, streams, in ponds, in fresh water, in saline water etc. Fresh and flowing water is very conducive to the crabs. Flowing condition of fresh water can be created artificially for the crab culture. They require highly oxygenated and bubling fresh water.

Crabs in nature breed and raise young ones and are very active during the rainy season when there is plenty of water flowing in the streams and rivers. Then they can be seen moving about on the rocks and edges of the streams and rivers.

All crabs have a hard shell on the dorsal side which protects the internal organs. This shell at the back side continues to the ventral side and form like a flap which ends into a pointed excretory outlet. Observing the ventral side one can distinguish between the male and female. Those with narrow nap are male and those with broad flap are females. On the edges of the flap of the female there are hair like structures called appendages. At the time of spawning the egg mass is held by these appendages.

Hence at the time of spawning a sticky spongy mass is seen held between the nap and the body but extruding outward. The eggs are hatched in about two weeks. In the sea crabs the young ones are like a larvae with a segmented body and tail. But in the land and fresh water crabs the young ones are like tiny adults. At the initial stage the young ones are held between the flap and the body of the mother till they are able to manage themselves without the mother. Crabs breed during rainy season. Artificial breeding is not yet known though it should be possible.

Practically very little information is available on crab culture. Yet they can be reared by breeding the selected males and females. Good care should be given to the young ones so that they survive till they are able to manage themselves.

Crabs can be reared in small tanks with shallow but clear, well oxygenated water flowing between the rocks which provides hiding place for the crabs. The tank may be as small as 4x4x4 feet, length, width and height. The size of the tank may be bigger according the land availability, size of the operation, number of crabs reared etc. Separate tanks for breeding stock, breeding and spawning, tanks for hatching, tanks for the young ones and tanks for the growers can be arranged. The requirement of tanks and other infrastructural facilities are arranged as per the need. Actually the rearer himself has to experiment on the system of crab rearing.

Crabs are omnivorous. Hence all kinds of feed materials and kitchen waste can be fed to them. The feed can be plant and animal material which can be collected and preserved as crab feed. Crabs are nocturnal in their habits. Still much observation and studies are required to develop a theory and practice of crab culture.

5. *Pearl culture*

Everyone is aware of what a pearl is and how valuable they are both in the national and world market. Pearls have been in usage for many years. They are concretion formed by a mollusc consisting of the same material called acre or mother-of-pearls as the molluscs shell. It is a highly valued gem- stone.

Pearls are characterized by their translucence and lustrons and a delicate play of surface colour called orient. More per- fect in shape are the spherical or drop like, and the deeper its lusture and the greater its value in the market. Only those pearls produced by molluscs whose shells are lined with "acre" (as in salt water and fresh water clams) are really the pearls. Pearls from other mollusks are reddish or whitish, porcelaneous or lacking in pearly lusture. Jewellers commonly refer to salt water pearls as oriental pearls and to those produced by fresh water molluscs as fresh water pearls.

The shell secreting cells of the mollusc are located in the mantle or epitheliums of its body. When a foreign particle penetrates the mantle the cells around the particles begin to secrete shell substance and cover it layer by layer in concentric circles forming the actual pearl in a period of 2-3 years.

Depending on the location of the foreign body there is qualitative difference in the pearls formed. When the particles are attached to the muscles of the mollusc irregularly shaped pearls are usually formed. Such pearls are called baroque pearls. Pearls that are grown close to the shell of mollusk are mostly flat on one side and round on the other side. They are called blister pearls. The spherical type of pearls are formed when the particle is located neither in the muscle nor close to the shell but somewhere in between which is called technically the nacre or mother of pearls.

The colour of the pearls vary from white to black. But the rose coloured Indian pearls are valued most. The size and weight of the pearls vary much the weight vary from 12.5 to 465 mg per pearl. The finest oriental pearls are produced by Pictada martensii species of salt water mollusc mainly found in Oman-Qatar areas of Persian Gulf, Manner areas between India and Sri Lanka, Indonesian islands and islands of South Pacific.

Pearling is a carefully fostered industry in central Europe. Mississippi river is famous for pearl production. Pearling has been going on in China since 1000 B.C. by intentionally inserting some foreign material into the molluscs. People in China used to catch molluscs, open them and insert into them mall mud pellets, tiny wood pieces, bone or small metal peices and leave them to the river or sea bed and wait for 3-4 year to pick them up to collect pearls from them. They were mostly producing blister pearls.

In 1890 Makinoto Kokichyi of Japan proved that introduction of a small bead into the mollusc at the correct position is a sure procedure for producing the pearls artificially. Since then Japan has been in fore front in the production of pearls. There are more than 2140 pearl farms in Japan. Following the example of Japan Australia also has become one of the leading pearl producing countries in the world.

6. *Turtle rearing*

Turtle is a collective name referring to Chelonia family. Turtles may be aquatic or terrestrial and the terretrial turtles are referred as tortoise though in common parlance no distinction is made between the turtle and tortoise.

Turtles are valued for their meat, egg and shell. Turtle meat is a delicacy; the eggs are big and nutritious and have medicinal value. The shell is very precious as they are very rare. The shells are used for making handicrafts. At present turtles are mostly caught from the rivers, sea and forest areas the demand for them do not match with the availability or supply. Hence attempts are already being made to rear them commercially. Turtles and tortoises can be easily reared by anyone though it is a long term investment.

Turtles are very slow growing but live for many years. They are very hardy and can withstand very long spell of dry and feed scarcity conditions. It can store water in the cloacal bladders like the camels storing water in their hump. It is an omnivorous though it prefers to be carnivorous. Hence feeding and watering turtles is not at all a problem for those who would like to rear them. Even if they are not fed for a number of days they will not die weekly or monthly feeding is enough.

The length of the turtles vary from few cm to 3. 7 metre and can weigh from few hundred grams to 680 kilograms.

Turtles lay eggs which are as good as any other kind of eggs used for human food. A female turtle lays on an average 300 eggs per year. But all the eggs are laid at one time on the land preferably in the sand. After laying the eggs the mother turtle has no role to play in hatching and in bringing up the young ones. One mating may be sufficient for a female to lay fertile eggs for several years. The number of days taken for incubation depends on the temperature.

The young one require practically no care either from the mother turtle nor from anyone. If they are provided sufficient feed and water and a rather congenial habitat they can be reared very successfully. The only negative point is that it will take many years to get some returns from the turtle farm. However the amount of money one gets per turtle is a justification enough for the undue delay of the returns.

No doubt we still lack sufficient scientific information about turtle and tortoise rearing under artificial conditions. Therefore one who rears the turtles and tortoises should collect as much information as possible before he starts the commercial rearing. After starting it he should go on acquiring information and knowledge both from outside as well as from his own experience. For this he should maintain records and accounts on various details observed during the turtle rearing. Keeping constant contact with people who study and rear the turtles and arranging platforms for mutual sharing and learning one can perfect the method of rearing turtles.

7. Crocodile rearing

Crocodiles are valued for eggs, meat and skin and has very high economic potential. They are reptiles belonging to the family Crocodilia and about 20 species of crocodiles have been reported so far. The term crocodile is a common name used for all the species in the Crocodilia family. Alligator is another term used for those crocodiles which have broad snout. Crocodiles are mostly found in the tropical areas of the world mainly in Atrica, America, India, Sri Lanka and South East Asian countries. They are nocturnal in habit and they remain in the water most of the time though they may crawl over to the land periodically.

Recently several crocodile farms have been started in many parts of the world. In India too there are a number of crocodile farms.

8. Aquatic weeds

Various types of aquatic weeds are edible to human beings and animals. Some of them are important raw material for a number industries and pharmaceutical firms. They may be fresh water or nonfresh water like the sea weeds. It is an established fact that much more vegetation is growing in the aquatic-ecosystem than in the terrestrial ecosystem. Hence in the future man will be compelled to look into the depth of sea and other aquatic ecosystems to satisfy their ever increasing demand for food and raw materials for various types of industries.

Of course cultivation of aquatic weeds is still an idea yet to be experimented and standardized. But as long as man is able to fulfill his requirements on the land he will not try for aquatic weed culture. The reason for not starting such ventures so far is that it is very expensive as well as it requires high level technology which is yet to be evolved and standardized.

V. Conclusion

Land being the limiting production factor, being already occupied and cultivated beyond the sustainability level and the land degradation being a serious environmental threat aquaculture offers to humanity a much more productive and profitable enterprise. India having more than five thousand kilometers of coastal line, numerous lakes, rivers and reservoirs hold much greater chance for establishing successful aquaculture systems for the future.

This is written with the intention of promoting ideas on various systems of aquaculture. No doubt some of them are very profitable enterprises if not all. And India has perhaps the best of the climatic and environmental conditions for a wide variety of aquacultural practices. It can even take a lead in this field. We need people who can take risks and forge ahead with new ideas and actions with research and learning mentality to evolve suitable practices in all the possible aquacultures.

3

MARINE FISHERIES ISSUES

Marine conservation, also known as marine resources conservation, is the protection and preservation of ecosystems in oceans and seas. Marine conservation focuses on limiting human-caused damage to marine ecosystems, and on restoring damaged marine ecosystems. Marine conservation also focuses on preserving vulnerable marine species.

Marine conservation is the study of conserving physical and biological marine resources and ecosystem functions. This is a relatively new discipline. Marine conservationists rely on a combination of scientific principles derived from marine biology, oceanography, and fisheries science, as well as on human factors such as demand for marine resources and marine law, economics and policy in order to determine how to best protect and conserve marine species and ecosystems. Marine conservation can be seen as a subdiscipline of conservation biology.

TECHNIQUES

Strategies and techniques for marine conservation tend to combine theoretical disciplines, such as population biology, with practical conservation strategies, such as setting up protected areas, as with marine protected areas (MPAs) or Voluntary Marine Conservation Areas. Other techniques include developing sustainable fisheries and restoring the populations of endangered species through artificial means.

Another focus of conservationists is on curtailing human activities that are detrimental to either marine ecosystems or species through policy, techniques such as fishing quotas, like those set up by the Northwest Atlantic Fisheries Organization, or laws such as those listed below. Recognizing the economics involved in human use of marine ecosystems is key, as is education of the public about conservation issues. This includes educating tourists that come to an area that might not be familiar of certain rules and regulations regarding the marine habitat. One example of this is a project called Green Fins that uses the SCUBA diving industry to educate the public based in SE Asia. This project, implemented by UNEP, encourages scuba diving operators to educate the public they teach to dive about the importance of marine conservation and encourage them to dive in an environmentally friendly manner that does not damage coral reefs or associated marine ecosystems.

Laws and treaties

International laws and treaties related to marine conservation include the 1966 Convention on Fishing and Conservation of Living Resources of the High Seas. United States laws related to marine conservation include the 1972 Marine Mammal Protection Act, as well as the 1972 Marine Protection, Research and Sanctuaries Act which established the National Marine Sanctuaries program.

In 2010, the Scottish Parliament enacted new legislation for the protection of marine life with the Marine (Scotland) Act 2010. The provisions in the Act include: Marine planning, Marine licensing, marine conservation, seal conservation, and enforcement.

Organizations and education

There are marine conservation organizations throughout the world that focus on funding conservation efforts, educating the public and stakeholders, and lobbying for conservation law and policy. Examples of these organizations are Oceana (non-profit group), the Marine Conservation Biology Institute (United States), Blue Frontier Campaign (United States), Frontier (the Society for Environmental Exploration) (United Kingdom), Marine Conservation Society (United Kingdom), The Reef-World Foundation (United Kingdom) and [Australian Marine Conservation Society].

On a regional level, PERSGA- the Regional Organization for the Conservation of the Environment of the Red Sea and the Gulf of Aden, is a regional entity serves as the secretariat for the Jeddah Convention-1982, one of the first regional marine agreements. PERSGA Member States are: Djibouti, Egypt, Jordan, Saudi Arabia, Somalia, Sudan and Yemen.

MARINE PROTECTED AREA

A Marine Protected Area (MPA) is a protected area whose boundaries include some area of ocean. "MPA" is often used as an umbrella term that describes marine areas that restrict human activity to protect living, non-living, cultural, and/or historic resources. Protections include limits on development, fishing gear types, fishing seasons, catch limits, moorings, to complete bans on removing marine life of any kind. As of 2010, the world hosted more than 5,000 MPAs, encompassing .8% of the ocean's surface. The world's MPAs are viewable in Google Earth.

Terminology

Marine Protected Area

Perhaps the simplest definition is any geographical area that includes some amount of ocean and that has specific limits on human activity for the purpose of protecting natural and/or cultural resources that do not apply to other nearby ocean areas. Various national, supra-national and other organizations offer alternatives that vary in scope and detail, but none of these are definitive. Most definitions require that the site must be set aside principally for conservation in order to be designated a Marine Protected Area. A site that is set aside, for example, for national defense which also has a local habitat will not qualify under the terms set by either IUCN or CBD as a Protected Area. Several types of compliant MPAs can be defined:

- A totally marine area with no significant terrestrial (land) parts.
- An area containing both marine and terrestrial components, which can vary between two extremes:
    - o A marine area that is mostly maritime, with little land; for example, an atoll would have a tiny island with a significant maritime population surrounding it.
    - o A marine area that is mostly terrestrial. In this case, whether or not it can be given such a title is largely debatable.
- Marine ecosystems that contain land and intertidal (land that is covered in/by water) components only. For example, a mangrove forest would contain no open sea or ocean marine environment, but its river-like marine ecosystem nevertheless constitutes under the definition.

The Convention on Biological Diversity attempted to solve this by defining the broader term of "Marine and Coastal Protected Area" (shorthand, MCPA):

Any defined area within or adjacent to the marine environment, together with its overlying water and associated flora, fauna, historical and cultural features, which has been reserved by legislation or other effective means, including custom, with the effect with the effect that its marine and/or coastal biodiversity enjoys a higher level of protection then its surroundings. et al. Barnard, p. 18

Some definitions require that the area be exclusively reserved for conservation, while others permit recreational and/or extractive activities. Others require that at least some part of the area lies below low tide, while others require only that it be at least near the shoreline.

The International Union for Conservation of Nature(IUCN) attempts to encompass these variations by defining six categories of protected area, based on management objectives and four broad governance types.

Marine Park

In Kenya Marine Parks prohibit fishing or extraction of resources of any kind, but allows recreation. In Tanzania, Marine Parks are zoned, and activities such as fishing are only allowed in certain, low risk areas.

Marine Reserve

In Kenya (and Belize), Marine Reserves allow for low-risk forms of fishing and are thus a lower area of protection. In Tanzania, Marine Reserves prohibit all removals. In scientific literature "Marine Reserve" the term usually connotes "maximum protection."

Areas, systems, networks, regions

While "area" refers to a single contiguous location, terms that group MPAs are not always consistently employed. These include "network", "system", and "region". In CBD decision VII/5, the agency agreed to the use of network on a global level, and the use of system on the national and regional level. The global level carries no authority or mandate, and all of the work lies in the system. The CBD defines the role of the global network as a large-scale network to be used as a mechanism to establish regional and local systems; some countries also define it based on their own terms.</ ref> Nations combine protected areas for purposes ranging from policy development and enforcement to marketing.

Other related terms

Related terms include Specially Protected Area (SPA), Special Area of Conservation (SAC), the United Kingdom's Marine Conservation Zones (MCZ) Marine reserve (MR), Marine park (MP), No Take Zone (NTZ), or Area of Special Conservation (ASC) Particularly Sensitive Sea Areas, Special Areas, etc., each have specific restrictions associated with them.

Usage restrictions

MPAs are established to protect a certain species, to benefit fisheries, rare habitat, or nursing grounds for fish or to protect entire ecosystems. MPAs are also established to protect historical sites such as shipwrecks and cultural sites such as aboriginal fishing grounds. MPAs can be very large (Great Barrier Reef) or very small (Area Marina Protetta Capo Rizzuto).

Typical restrictions in MPAs include ones on fishing, oil and gas mining and tourism. Other restrictions may limit the use of ultrasonic devices like SONAR (which may confuse the guidance system of cetaceans), development, construction and the like. Still others, such as New Zealand's marine reserves, are 'no take' areas, where all forms of exploitation are prohibited. Even ship transit can be regulated or banned, either as a preventive measure or to avoid direct disturbance to certain species. The degree to which environmental regulations affect shipping varies according to whether MPAs are located in territorial waters, exclusive economic zones, or the high seas. The law of the sea regulates these limits.

For this reason, most MPAs have been located in territorial waters, where enforcement can be ensured. However, MPAs can also be established in a state's exclusive economic zone and even international waters. For example, Italy, France and Monaco in 1999 jointly established a cetacean sanctuary in the Ligurian Sea named the Pelagos Sanctuary for Mediterranean Marine Mammals. This sanctuary includes both national and international waters.

Both the CBD and IUCN recommend that a variety of possible management systems be considered when designing a protected area system. They advocate that MPAs be seen as one of many "nodes" in a network of protected areas. The following are the most commonly used individual types of MPAs.

No-take areas

The highest degree of protection is the no-take area, which severely limits human activities. Generally, they prohibit removing anything from the protected area. No-take can cover the whole MPA, or specific vulnerable portions that enjoy elevated protection.

The IUCN definition allows the extraction of resources from the area only with a permit and for scientific use only. There is no globally-accepted definition.

Seasonal and temporary management

Activities, most critically fishing, are restricted seasonally or temporarily to let the area recover. The most common use seasonal limits to protect fish populations during vital periods, such as spawning season. The "Irish Sea Cod Box" is such a season. Another use is to temporarily protect a depleted marine population from overfishing, allowing it to recover, as in the waters of Okinawa, Japan.

Multiple-use

Increasingly, multiple use MPAs are the most common and arguably most effective type. These areas employ two or more types. This flexibility allows the most important sections get the highest protection. A common practice is to make the most critical area fully no-take, surrounding it with areas of lesser protections. The island of Asinara is an example of such an MPA.

Management

The two families of approaches for managing MPAs are community-managed and politically-managed.

Community-managed and related approaches

Community-managed MPAs empower local communities to manage marine resources partially or completely independent of the governmental jurisdictions they inhabit. They are not always officially recognized, depending on the political environment. Empowering communities to manage resources can lower conflict levels and help fisheries recover. This approach can provide direct influence for all involved, including subsistence and commercial fishers, scientists, tourism businesses, youths and others. They often fall into the following, unrelated, designations (although there is overlap):

- World Heritage Site (WHS) - an area exhibiting extensive natural or cultural history. Maritime areas are poorly represented, however, with only 31 out of over 800 sites. One example of overlap is the island of Asinara.
- Man and the Biosphere - This UNESCO program promotes "a balanced relationship between humans and the biosphere." Under article 4, biosphere reserves must "encompass a mosaic of ecological systems", and thus combine terrestrial, coastal, or marine ecosystems. In structure they are similar to Multiple-use MPAs, with a core area ringed by different degrees of protection.
- Ramsar Site - These sites must meet certain criteria for the definition of "Wetland" to become part of a global system. These sites do not necessarily receive protection, but are indexed by importance for later recommendation to an agency that could designate it a protected area.

Fishery management areas

Areas managed only to sustain fisheries occasionally change to become MPAs. One example is the Fish Habitat Reserves in Australia.

International efforts

Historically, Marine Protected areas have been established on an ad hoc basis by individual nations. The 17th International Union for Conservation of Nature (IUCN) General Assembly in San Jose, California, the 19th IUCN assembly and the fourth World Parks Congress all proposed to centralize the activity. The World Summit on Sustainable Development in 2002 called for the establishment of marine protected areas consistent with international laws and based on scientific information, including representative networks by 2012.

The Evian agreement, signed by G8 Nations in 2003, agreed to these terms. The Durban Action Plan, developed in 2003, called for regional action and targets to establish a network of protected areas by 2010 within the jurisdiction of regional environmental protocols. It recommended establishing protected areas for 20 to 30% of the world's oceans by the goal date of 2012. The Convention on Biological Diversity considered these recommendations and recommended requiring nations to set up marine parks that are controlled by a central organization before merging them. The United Nations Framework Convention on Climate Change agreed to the terms laid out by the convention, and its member nations committed to the target in 2004, signing the statement at right. United Nations Decision VII/28 of the laid out the following deadlines:

- By 2006 complete area system gap analysis at national and regional levels.
- By 2008 address the underrepresented of marine ecosystems in existing national and regional systems of protected areas, taking in account marine ecosystems beyond areas of national jurisdiction in accordance with applicable international laws.
- By 2009 designate the protected areas identified through the gap analysis.
- By 2012 complete the establishment of a comprehensive and ecologically representative national and regional system of Marine Protected Areas.

The establishment by 2010 of terrestrial and by 2012 for marine areas of comprehensive, effectively managed, and ecologically representative national and regional systems of protected areas that collectively, inter alia through a global network, contribute to achieving the three objectives of the Convention and the 2010 target to significantly reduce the current late of biodiversity loss at the global, regional, national, and sub-national levels and contribute to poverty reduction and the pursuit of sustainable development.

The UN later also endorsed another decision, Decision VII/15, in 2006:

Effective conservation of 10% of each of the world's ecological regions by 2010.

- United Nations Framework Convention on Climate Change Decision VII/15

Many countries have established national targets, accompanied by action plans and implementations. The UN Council identified the need for countries to collaborate with each other to establish effective regional conservation plans. A few of these national targets are listed in the table below

Country	Plan of Action
American Samoa	20% of reefs to be protected by 2010
South Australia	19 Marine Protected Areas by 2010
Bahamas	20% of the marine ecosystem protected for fishery replenishment by 2010.
	20% of coastal and marine habitats by 2015.

Belize	20% of bioregions.
	30% of Coral reefs.
	60% of turtle nesting sites.
	30% of Manatee distribution.
	60% of American crocodile nesting.
	80% of Breeding areas.
Chile	10% of marine areas by 2010. National network for organization by 2015.
Cuba	22% of land habitat, including:
	15% of the Insular shelf
	25% of Coral reefs
	25% of Wetlands
Dominican Republic	20% of Marine and Coastal by 2020.
Micronesia	30% of shoreline ecosystems by 2020.
Fiji	30% of reefs by 2015.
	30% of water managed by Marine Protected Areas by 2020.
Germany	38% of water managed by the Marine Protected network. (no set date)
Grenada	25% of nearby marine resources by 2020.
Guam	30% of nearby marine ecosystem by 2020.
Indonesia	100,000 km2 by 2010. 200,000 km2 by 2020.
Jamaica	20% of marine habitats by 2020.
Madagascar	100,000 km2 by 2012.
Marshal Islands	30% of nearby marine ecosystem by 2020.
New Zealand	20% of marine environment by 2010.
North Mariana Islands	30% of nearby marine ecosystem by 2020.
Palau	30% of nearby marine ecosystem by 2020.
Peru	Marine Protected Area system established by 2015.
Philippines	10% Fully Protected by 2020.
Senegal	Creation of MPA network. (no set date)
St. Vincent and the Grenadines	20% of marine areas by 2020.
Tanzania	10% of marine area by 2010; 20% by 2020.

United Kingdom	Establish an ecologically coherent network of marine protected areas by 2012.
USA - California	29 MPAs covering 18% of state marine area with 243 square kilometres (94 sq mi) at maximum protection.

In 1981 the World Conservation Monitoring Centre began compiling the World Database on Protected Areas, tracking information related to Protected areas (PAs) from governmental, private, and scientific work. In 2005, an online database named "MPAGlobal" was established to better organize information related specifically to marine protected areas. This was fully reintegrated into the original system in late 2008.

Organizing principles

Global-UNEP-RSP

The United Nations Environmental Program arranges MPAs in a global program called UNEP-RSP (Regional Seas Program), comprising thirteen regions and five partner programs. Participants are linked either through a convention or a regional program. The five independent partner programs are active, but not under the UNEP jurisdiction. The arrangement is based on biology and geography rather than national or other political divisions.

The marine environment also benefits more than land areas from systematic protection because underwater, national borders have no physical presence. Water, heat, waves, and animals move across them with few or no restrictions.

Local networks

Local MPA networks are usually built in one of two ways.

- Biologists argue for designation to preserve biodiversity (usually an endangered species). Their request is adjusted by stakeholders until it is agreed upon or rejected. This was the first, and previously the most common, approach.
- Under systematic conservation planning, network organization flows from understanding the interaction between species in one MPA with those in another.

United States

The U.S. national system of MPAs includes an assemblage of sites, systems, and networks established and managed by federal, state, tribal, or local governments that work together to conserve important natural and cultural resources. Although each MPA is independently managed, the national system provides opportunities for cooperation, and promotes public participation in MPA decision-making by improving access to scientific and public policy information.

MPAs join the national system via a process that is designed to be transparent and science-based, with opportunity for public comment.

The entry criteria are:

- Meets the definition of an MPA
- Has a management plan for the specific site
- Contributes to at least one priority conservation objective as listed in the Framework
- Cultural heritage MPAs must also conform to criteria for the National Register of Historic Places

The nomination process involves:

- The MPA Center uses the MPA Inventory to identify potentially eligible MPAs, and invites them to nominate their sites.
- Managing entities submit a nomination form to the MPA Center.
- Nominated sites are published in a Federal Register notice and posted on www.mpa.gov for a 30-day public comment period.
- Based on public comment, managing entities reaffirm or withdraw their nominations. * After final MPA Center review, remaining MPAs join the national system. Accepted MPAs are listed in the official List of National System MPAs, and are published in the Federal Register and on www.mpa.gov.

Global status

Greater Caribbean

The Greater Caribbean subdivision encompasses an area of about 5,700,000 square kilometres (2,200,000 sq mi) of ocean and 38 very diverse nations. The area includes island countries like the Bahamas and Cuba, and the majority of Central America.

The Convention for Protection and Development of the Marine Environment of the Wider Caribbean Region (better known as the Cartagena Convention) was established in 1983, and protocols involving protected areas were ratified in 1990. As of 2008, there are about 500 MPAs in the region. Coral reefs are the best represented.

Two networks are under development, the Mesoamerican Barrier Reef System (a long barrier reef that borders the coast of much of Central America), and the "Islands in the Stream" program (covering the Gulf of Mexico).

Latin America

Latin America in particular considers itself one large MPA system. As of 2008, 0.5% of the Latin American marine environment is protected, mostly through the use of small, multiple-use MPAs

South Pacific

The South Pacific network ranges from Belize to Chile. Governments in the region adopted the Lima Convention and Action Plan for protected areas for the South Pacific region in 1981; an MPA-specific protocol was ratified in 1989. The Permanent Commission on the Exploitation and Conservation on the Marine Resources of the South Pacific promotes the exchange of studies and information among participants.

The region is currently running one comprehensive cross-national program, the Tropical Eastern Pacific Marine Corridor Network, signed in April 2004. The network covers about 211,000,000 square kilometres (81,000,000 sq mi). The participating countries are Panama, Costa Rica, Colombia, and Ecuador.

North Pacific

The North Pacific network covers the western coasts of Mexico, Canada, and the U.S. The "Antigua Convention" and an action plan for the north Pacific region were adapted in 2002. There, is however, no protocol; participant nations manage their own national systems. In 2010-2011, the State of California is completing a series of hearings and actions via the state Department of Fish and Game to establish new MPA's. Although highly controversial among fishing circles, the MPA's are going forward with decisions still pending as to where and how large.

United States

In April 2009, the United States established a National System of Marine Protected Areas, which strengthens the protection of U.S. ocean, coastal, and Great Lakes resources. As of 2009, 225 MPAs participated in the national system. Sites agree to work together toward common national and regional conservation goals and priorities. NOAA's National Marine Protected Area's Center maintains a comprehensive inventory of all 1,600+ MPAs within the Exclusive Economic Zone of the United States. Most MPAs in the U.S.

allow some type of extractive use. Less than 1% of U.S. waters prohibit all extractive activities.

Europe

The Natura 2000 ecological network of pr.otected areas in the territory of the European Union including a wide range of MPA in the Nortatlantic, the Mediterranian Sea and the Baltic Sea. The member-staates has to define NATURA 2000-areas at sea in their Exclusive Economic Zone.

Non-governmental organizations

In 1999, the North American Marine Protected Areas Network (NAMPAN) was established, a project of the Commission for Environmental Cooperation, (CEC). NAMPAN is a virtual network of people and places spanning Canada, The United States, and Mexico that work together to address tri-national, transboundary marine conservation issues. Currently Canada and the United States are developing country-wide systems of marine protected areas.

The Commission for Environmental Cooperation, established in 1999, coordinates activities, with representatives from Mexico, the USA, and Canada. There are currently two cross-national networks in development. In 2005, the Commission proposed the Baja California to Bering Sea (B2B) initiative, involving 28 areas.

Noteworthy Sites

As of February 2009, there were approximately 5,000 MPAs around the world, covering 0.8% of the world's oceans.

Notable MPAs include:

- The Bowie Seamount on the Coast of British Columbia, Canada.
- The Great Barrier Reef in Queensland, Australia.
- The Ligurian Sea Cetacean Sanctuary in the seas of Italy, Monaco and France.
- The Dry Tortugas in the Florida Keys, USA.
- The Papah?naumoku?kea Marine National Monument in Hawaii.
- The Phoenix Islands Protected Area, Kiribati.
- The Channel Islands Marine Protected Areas in California, USA.

Effectiveness

Criteria

Both CBD and IUCN have criteria for setting up and maintaining MPA networks, which emphasize 4 factors:

- Adequacy-ensuring that the sites have the size, shape, and distribution to ensure the success of selected species.
- Representability-protection for all of the local environment's biological processes.
- Resilience-the resistance of the system to natural disaster, such as a tsunami or flood.
- Connectivity-maintaining population links across nearby MPAs.

A learning model of migratory fish behavior and fishing interaction predicted that closed areas (MPAs) would increase fish biomass and decrease fish catches, that closing spawning areas to fishers would increase mean fish biomass, with lower variance, but that without catch restrictions throughout the range of the fish, long-term fish biomass and resulting catches would still decrease. The model also predicts that higher fish mobility also would increase fish biomass, but decrease fish catches.

Managers and scientists use geographic information systems and remote sensing to map and analyze MPAs. NOAA Coastal Services Center compiled an "Inventory of GIS-Based Decision-Support Tools for MPAs." The report focuses on GIS tools with the highest utility for MPA processes. Remote sensing uses advances in aerial photography image capture, satellite imagery, acoustic data, and radar imagery.

Protecting Red Coral

Two assessments, conducted thirty years apart, of three Mediterranean MPAs demonstrate that proper protection allows commercially valuable and slow-growing red coral (Corallium rubrum) to produce large colonies in shallow water of less than 50 metres (160 ft). Shallow-water colonies outside these decades-old MPAs are typically very small. The MPAs are Banyuls, Carry-le-Rouet and Scandola, off the island of Corsica.

Criticism

Some existing and proposed MPAs have been criticized by local indigenous populations, and their supporters, as impinging on land usage rights. One example of this is the proposed Chagos Protected Area in the Chagos Islands, contested by Chagossians deported from their homeland in 1965 by the British in the creation of the British Indian Ocean Territory.

According to Wikileaks CableGate documents, the UK proposed that the BIOT become a "marine reserve" with the aim of preventing the former inhabitents from returning to their lands and to protect the joint United

Kingdom and United States military base on Diego Garcia Island. One alternative to imposing MPAs on an indigenous population is through the use of Indigenous Protected Areas, such as those in Australia.

MARINE POLLUTION

Marine pollution occurs when harmful effects, or potentially harmful effects, can result from the entry into the ocean of chemicals, particles, industrial, agricultural and residential waste, noise, or the spread of invasive organisms. Most sources of marine pollution are land based. The pollution often comes from nonpoint sources such as agricultural runoff and wind blown debris. Many potentially toxic chemicals adhere to tiny particles which are then taken up by plankton and benthos animals, most of which are either deposit or filter feeders. In this way, the toxins are concentrated upward within ocean food chains. Many particles combine chemically in a manner highly depletive of oxygen, causing estuaries to become anoxic.

When pesticides are incorporated into the marine ecosystem, they quickly become absorbed into marine food webs. Once in the food webs, these pesticides can cause mutations, as well as diseases, which can be harmful to humans as well as the entire food web.

Toxic metals can also be introduced into marine food webs. These can cause a change to tissue matter, biochemistry, behaviour, reproduction, and suppress growth in marine life. Also, many animal feeds have a high fish meal or fish hydrolysate content. In this way, marine toxins can be transferred to land animals, and appear later in meat and dairy products.

Although marine pollution has a long history, significant international laws to counter it were enacted in the twentieth century. Marine pollution was a concern during several United Nations Conferences on the Law of the Sea beginning in the 1950s. Most scientists believed that the oceans were so vast that they had unlimited ability to dilute, and thus render harmless, pollution. In the late 1950s and early 1960s, there were several controversies about dumping radioactive waste off the coasts of the United States by companies licensed by the Atomic Energy Commission, into the Irish Sea from the British reprocessing facility at Windscale, and into the Mediterranean Sea by the French Commissariat à l'Energie Atomique. After the Mediterranean Sea controversy, for example, Jacques Cousteau became a worldwide figure in the campaign to stop marine pollution. Marine pollution made further international headlines after the 1967 crash of the oil tanker Torrey Canyon, and after the 1969 Santa Barbara oil spill off the coast of California. Marine pollution was a major area of discussion during the 1972 United Nations Conference on the Human Environment, held in Stockholm. That year also saw the signing of the Convention on the Prevention of Marine Pollution by Dumping of Wastes and Other Matter, sometimes called the London Convention. The London Convention did not ban marine pollution, but it established black and gray lists for substances to be banned (black) or regulated by national authorities (gray). Cyanide and high-level radioactive waste, for example, were put on the black list. The London Convention applied only to waste dumped from ships, and thus did nothing to regulate waste discharged as liquids from pipelines.

There are many different ways to categorize, and examine the inputs of pollution into our marine ecosystems. Patin (n.d.) notes that generally there are three main types of inputs of pollution into the ocean: direct discharge of waste into the oceans, runoff into the waters due to rain, and pollutants that are released from the atmosphere.

One common path of entry by contaminants to the sea are rivers. The evaporation of water from oceans exceeds precipitation. The balance is restored by rain over the continents entering rivers and then being returned to the sea. The Hudson in New York State and the Raritan in New Jersey, which empty at the northern and southern ends of Staten Island, are a source of mercury contamination of zooplankton (copepods) in the open ocean. The highest concentration in the filter-feeding copepods is not at the mouths of these rivers but 70 miles south, nearer Atlantic City, because water flows close to the coast. It takes a few days before toxins are taken up by the plankton.

Pollution is often classed as point source or nonpoint source pollution. Point source pollution occurs when there is a single, identifiable, and localized source of the pollution. An example is directly discharging sewage and industrial waste into the ocean. Pollution such as this occurs particularly in developing nations. Nonpoint source pollution occurs when the pollution comes from ill-defined and diffuse sources. These can be difficult to regulate. Agricultural runoff and wind blown debris are prime examples.

MERCURY IN FISH

Fish and shellfish concentrate mercury in their bodies, often in the form of methylmercury, a highly toxic organic compound of mercury. Fish products have been shown to contain varying amounts of heavy metals, particularly mercury and fat-soluble pollutants from water pollution. Species of fish that are long-lived and high on the food chain, such as marlin, tuna, shark, swordfish, king mackerel, tilefish, northern pike, and lake trout contain higher concentrations of mercury than others. The presence of mercury in fish can be a health issue, particularly for women who are or may become pregnant, nursing mothers, and young children.

Biomagnification

Mercury and methylmercury is present in only small concentrations in seawater. However, it is absorbed, usually as methylmercury, by algae at the start of the food chain. This algae is then eaten by fish and other organisms higher in the food chain. Fish efficiently absorb methylmercury, but only very slowly excrete it. Methylmercury is not soluble and therefore is not apt to be excreted. Instead, it accumulates, primarily in the viscera although also in the muscle tissue. This results in the bioaccumulation of mercury, in a buildup in the adipose tissue of successive trophic levels: zooplankton, small nekton, larger fish etc. Anything which eats these fish within the food chain also consumes the higher level of mercury the fish have accumulated. This process explains why predatory fish such as swordfish and sharks or birds like osprey and eagles have higher concentrations of mercury in their tissue than could be accounted for by direct exposure alone. Species on the food chain can amass body concentrations of mercury up to ten times higher than the species they consume. This process is called biomagnification. For example, herring contains mercury levels at about 0.01 ppm while shark contains mercury levels greater than 1 ppm.

Range of Contamination

U.S. government scientists tested fish in 291 streams around the country for mercury contamination. They found mercury in every fish tested, according to the study by the U.S. Department of the Interior. They found mercury even in fish of isolated rural waterways. Twenty five percent of the fish tested had mercury levels above the safety levels determined by the U.S. Environmental Protection Agency for people who eat the fish regularly.

Sources

Much of the mercury that eventually finds its way into fish originates with coal-burning power plants and chlorine production plants. The largest source of mercury contamination in the United States is coal-fueled power plant emissions. Chlorine chemical plants use mercury to extract chlorine from salt, which in many parts of the world is discharged as mercury compounds in waste water, though this process has been replaced for the most part by the more economically viable membrane cell process, which does not use mercury. Coal contains mercury as a natural contaminant. When it is fired for electricity generation, the mercury is released as smoke into the atmosphere. Most of this mercury pollution can be eliminated if pollution-control devices are installed.

Current Advice

The complexities associated with mercury transport and environmental fate are described by USEPA in their 1997 Mercury Study Report to Congress.

Because methylmercury and high levels of elemental mercury can be particularly toxic to a fetus or young children, organizations such as the U.S. EPA and FDA recommend that women who are pregnant or plan to become pregnant within the next one or two years, as well as young children avoid eating more than 6 ounces (one average meal) of fish per week. In the United States, the FDA has an action level for methylmercury in commercial marine and freshwater fish that is 1.0 parts per million (ppm). In Canada, the limit for the total of mercury content is 0.5 ppm. The Got Mercury? website includes a calculator for determining mercury levels in fish.

Species with characteristically low levels of mercury include shrimp, tilapia, salmon, pollock, and catfish (FDA March 2004). The FDA characterizes shrimp, catfish, pollock, salmon, sardines, and canned light tuna as low-mercury seafood, although recent tests have indicated that up to 6 percent of canned light tuna may contain high levels. A study published in 2008 found that mercury distribution in tuna meat is inversely related to the lipid content, suggesting that the lipid concentration within edible tuna tissues has a diluting effect on mercury content. These findings suggest that choosing to consume a type of tuna that has a higher natural fat content may help reduce the amount of mercury intake, compared to consuming tuna with a low fat content. Also, many of the fish chosen for sushi contain high levels of mercury.

According to the US Food and Drug Administration (FDA), the risk from mercury by eating fish and shellfish shall not be a health concern for most people. However, certain seafood might contain levels of mercury that may cause harm to an unborn baby (and especially its brain development and nervous system). In a young child, high levels of mercury can interfere with the development of the nervous system. The FDA provides three recommendations for young children, pregnant women, and women of child-bearing age:

1. Do not eat shark, swordfish, king mackerel, or tilefish because they might contain high levels of mercury.
2. Eat up to 12 ounces (2 average meals) a week of a variety of fish and shellfish that are lower in mercury. Five of the most commonly eaten fish and shellfish that are low in mercury are: shrimp, canned light tuna, salmon, pollock, and catfish. Another commonly eaten fish, albacore or big eye ("white") tuna depending on its origin might have more mercury than canned light tuna. So, when choosing your two meals of fish and shellfish, it is recommended that you should not eat more than up to 6 ounces (one average meal) of albacore tuna per week.
3. Check local advisories about the safety of fish caught by family and friends in your local lakes, rivers, and coastal areas. If no advice is

available, eat up to 6 ounces (one average meal) per week of fish you catch from local waters, but consume no other fish during that week.

Background

In 1950s, inhabitants of the seaside town of Minamata, on Kyushu island in Japan, noticed strange behavior in animals. Cats would exhibit nervous tremors, dance and scream. Within a few years this was observed in other animals; birds would drop out of the sky. Symptoms were also observed in fish, an important component of diet, especially for the poor. When human symptoms started to be noticed around 1956 an investigation began. Fishing was officially banned in 1957. It was found that the Chisso Corporation, a petrochemical company and maker of plastics such as vinyl chloride, had been discharging heavy metal waste into the sea. They used mercury compounds as catalysts in their syntheses. It is believed that about 5,000 people were killed and perhaps 50,000 have been to some extent poisoned by mercury. Mercury poisoning in Minamata, Japan, is now known as Minamata disease. A book by Jane Hightower, Diagnosis Mercury: Money, Politics and Poison, published in 2008, is largely about humans exposed to mercury through eating large predatory fish such as swordfish, shark, king mackerel, large tuna, etc.

SHARK FINNING

Shark finning refers to the removal and retention of shark fins and the discarding at sea of the rest of the fish. Shark finning takes place at sea so the fishers only have to transport the fins. Shark finning is widespread, and largely unmanaged and unmonitored. Shark finning has increased over the past decade largely due to the increasing demand for shark fins for Chinese shark fin soup and traditional cures, improved fishing technology, and improved market economics.

Some researchers reckoned that from 1996 to 2000, 26 to 73 million sharks were traded yearly. The annual median for the period was 38 million, nearly four times more than the UN estimates but considerably lower than those of many conservationists. Shark fins are a billion dollar industry.

According to wildlife conservation much of the sharks' fin trade uses fins cut from living sharks, called finning. Because shark meat is worth much less, the now finless and often still-living sharks are thrown back into the sea to make room for more of the valuable fins. In the ocean, the sharks either die from suffocation or are eaten because they are unable to move normally.

Impact and Reporting

According to Giam Choo Hoo, the longest serving member of The Convention on International Trade in Endangered Species of Wild Fauna and Flora Animals Committee, "The perception that it is common practice to kill sharks for only their fins - and to cut them off whilst the sharks are still alive - is wrong.... The vast majority of fins in the market are taken from sharks after their death." However, some researchers dispute this claim after extensive examination of fin sourcing and fisheries data; one study of sharks harvested for their valuable fins estimates that between 26-73 million sharks are killed each year worldwide, which is almost three times higher than official Food and Agriculture Organization estimates.

The crew of the conservation vessel Ocean Warrior witnessed and photographed industrial-scale finning within Costa Rica's Cocos Island National Park protected marine area. The practice is featured in the documentary Sharks: Stewards of the Reef, which contains footage from Western Australia and Central America and also examines shark finning's cultural, financial and ecological impacts. Underwater photographer Richard Merritt also has witnessed finning of living sharks in Indonesia where he saw immobile finless sharks lying on the sea bed still alive below the fishing boat. Finning has been witnessed and filmed within a protected marine area in the Raja Ampat islands of Indonesia.

Animal welfare groups vigorously oppose finning on moral grounds and also because it is one cause for the rapid decline of global shark populations. On the IUCN red list there are 39 species of elasmobranches (sharks and rays) listed as threatened species (Critically Endangered, Endangered or Vulnerable). CITES lists three sharks in Appendix II: the basking shark, the great white shark, and the whale shark. Appendix II lists those species that are not in danger of extinction, but which require controls on international trade to maintain their populations. It is estimated that 10-100 million sharks are slaughtered each year for their fins with a median figure of 38 million. The industry is valued at US$1.2 billion; because of the lucrative profits, there are allegations of links to organized crime. They also raise questions on the medical harm from the consumption of high levels of toxic mercury reportedly found in shark fins.

Numbers of some shark species have dropped as much as 80% over the last 50 years. Some organizations claim that shark fishing or bycatch (the unintentional capture of species by other fisheries) is the reason for the decline in some species' populations and that the market for fins has very little impact - bycatch accounts for an estimated 50% of all sharks taken - others that the market for shark fin soup is the main reason for the decline. Tommy Cheung, the legislator representing Hong Kong's catering sector, said: "I don't believe sharks are an endangered species. Some species of shark may be, but not all shark's fin comes from certain species. There are a lot of species that are plentiful." There is no reliable count for the numbers taken in the shark fin trade and thus it is hard to prove the claims on either side of the argument.

According to Giam's article, "sharks are caught virtually all parts of the world. Despite the strongly declared objectives of the Fisheries Commission in Brussels, there are very few restrictions on fishing for sharks in European waters. The meat of dogfishes, smoothhounds, cat sharks, skates and rays is in high demand by European consumers...The situation in Canada and the United States is similar: the blue shark is sought after as a sport fish while the porbeagle, mako and spiny dogfish are part of the commercial fishery...the truth is this: Sharks will continue to be caught and killed on a wide scale by the more organized and sophisticated fishing nations... targeting shark's fin soup will not stop this accidental catch. The fins from these catches will be thrown away or turned into animal feed and fertilizers if shark's fin soup is shunned." In March 2011, the VOA Special English service of the Voice of America broadcast a 15-minute science program on shark finning.

International reaction

Many countries now prohibit finning; however, many international waters are unregulated. International fishing authorities are considering banning shark fishing (and finning) in the Atlantic Ocean and Mediterranean Sea. Finning is banned in the Eastern Pacific, but shark fishing and finning continues unabated in most of the Pacific and Indian Ocean. In countries such as Thailand and Singapore, public awareness advertisements on finning have reportedly reduced consumption by 25%. In 2007, Canadian filmmaker and biologist Rob Stewart created a film, Sharkwater, which exposes the shark fin industry in detail.

China

NBA All-Star Yao Ming pledged to stop eating shark fin soup at a news conference on August 2, 2006. Yao's comments were largely unreported in the Chinese media and drew a reproach from Chinese seafood industry associations. Ironically, shark fin soup was on Yao Ming's wedding dinner menu. U.S. basketball player Tracy McGrady, a team mate of Yao's, reportedly said that he was impressed by the soup when he tried it for the first time, but was criticized by the Hong Kong branch of the World Wide Fund for Nature for his remark. Late Australian naturalist Steve Irwin was known to walk out of Chinese restaurants if he saw shark fin soup on the menu. The Chinese-American chef, Ken Hom, sees the West doing little to protect stocks of cod and caviar-producing sturgeon despite the outcry over shark-finning, but he also stresses the wastefulness of harvesting only the fins.

Hong Kong

Hong Kong Disneyland dropped shark fin soup from its wedding banquet menu after international pressure from environmental groups, who threatened to boycott its parks worldwide despite the high demand for the delicacy in China. The University of Hong Kong has banned shark fin on campus.

Malaysia

On September 15, 2007, Malaysia's Natural Resources and Environment Ministry Azmi Khalid banned shark's fin soup from official functions committing to the Malaysian Nature Society (for conservation of shark species).

New Zealand

The great white sharks have been given full protection in the territorial waters of New Zealand but shark finning is legal on other shark species if the shark is dead. The Royal Forest and Bird Protection Society of New Zealand are campaigning to raise awareness of shark finning and a number of foodies have fronted the campaign.

Palau

In 2009, the Republic of Palau created the world's first shark sanctuary. It is illegal to catch sharks within Palau's EEZ, which covers an area of 230,000 square miles (600,000 km2). This is an area about the size of France. President Johnson Toribiong also called for a ban on global shark finning, stating: "These creatures are being slaughtered and are perhaps at the brink of extinction unless we take positive action to protect them."

United States

Bill Clinton banned finning in the United States and with U.S.-registered vessels, but not foreign-registered vessels. Shark fins cannot be imported into the U.S. without the associated carcass. In 1991, the percentage of sharks killed by U.S. longline fisheries in the Pacific Ocean for finning was approximately 3%. By 1998, that percentage had grown to 60%. Between 1991 and 1998, the number of sharks retained by the Hawaii-based swordfish and tuna longline fishery had increased from 2,289 to 60,857 annually, and by 1998, an estimated 98% of these sharks were killed for their fins.

In 2010, Hawaii became the first state to ban the possession, sale and distribution of shark fins. The law will take effect on July 1, 2011. In January 2011, President Barack Obama signed the Shark Conservation Act into law to close the loopholes of the 2000 Shark Finning Prohibition Act. Specifically, the new law prohibits any boat to carry shark fins without the corresponding number and weight of carcasses, and all sharks must be brought to port with their fins attached.

4

MARINE FISHERIES ORGANIZATION

NATIONAL MARINE FISHERIES SERVICE

The National Marine Fisheries Service (NMFS) is a United States federal agency. A division of the National Oceanic and Atmospheric Administration (NOAA) and the Department of Commerce, NMFS is responsible for the stewardship and management of the nation's living marine resources and their habitat within the United States' Exclusive Economic Zone, which extends seaward 200 nautical miles from the coastline (about 370 Kilometers).

In recent years, the agency has come under intense scrutiny from the fishing industry, both commercial and recreational, and Congress leading to a multi-part investigation by the Commerce Department Inspector General which found serious problems and misuse of funds.

During her confirmation hearing, Dr. Jane Lubchenco, NOAA Administrator in the Obama Administration, expressed a commitment to fix a relationship between the fishing industry and NMFS. Lubchenco herself called the relationship between NMFS and those who it regulates "seriously dysfunctional." NOAA oversees the NMFS.

Using the tools provided by the Magnuson-Stevens Fishery Conservation and Management Act, the NMFS assesses and predicts the status of fish stocks, ensures compliance with fisheries regulations and works to end wasteful fishing practices. Under the Marine Mammal Protection Act and the Endangered Species Act, the agency is also tasked with recovering protected marine species such as wild salmon, whales and sea turtles.

With the help of the six regional science centers, eight regional fisheries management councils, the coastal states and territories, and three interstate fisheries management commissions, NMFS conserves and manages marine fisheries to promote sustainability and to prevent lost economic potential associated with overfishing, declining species, and degraded habitats. While the coastal states and territories generally have authority to manage fisheries within near-shore state waters, the NMFS has the primary responsibility to conserve and manage marine fisheries in the U.S. exclusive economic zone beyond state waters. The agency also attempts to balance competing public needs for the natural resources under its management.

Law Enforcement

The NMFS also serves as a federal law enforcement agency, working closely with state enforcement agencies, the U.S. Coast Guard, and foreign enforcement authorities. The National Oceanic and Atmospheric Administration Fisheries Office for Law Enforcement is based in Silver Spring, Maryland.

Dale Jones, former director of federal fishing law enforcement, and his deputy director Mark Spurrier were removed from their positions in April of 2010 of after a series of audits by U.S. Department of Commerce Inspector General Todd Zinser. Mr. Zinser and his investigators found that Jones had presided over wildly excessive and disproportionate treatment of New England fishermen, misuse of government funds, and a document shredding party while Zinser was actively investigating his department. Andrew Cohen, Special Agent in Charge NER, was relieved of his position in September 2010 due to findings of overzealous law enforcement practices.

During 2010, the U.S. Department of Commerce Office of Inspector General issued four Reports or Reviews of NOAA Fisheries Enforcement Programs and Operations. Here are the links to the four reports January 2010 OIG Report, April 2010 OIG Report on Destruction of Documents at NOAA Fisheries During an Ongoing OIG Review, July 2010 OIG Report on Asset Forfeiture Fund, September 2010 Final OIG Report. The cluster of OIG Reports found such a troubling situation within NOAA that on September 23, 2010 Secretary of Commerce Gary Locke reached down into NOAA and took command himself for resolving the issues raised by Inspector General Zinser Secretary Locke to Appoint Special Master to Review NOAA Law Enforcement Cases, Restricts Use of the Asset Forfeiture Fund.

Regulations

The NMFS regulatory program is one of the most active in the Federal government, with hundreds of regulations published annually in the Federal Register. Most regulations are published to conserve marine fisheries under the Magnuson-Stevens Fishery Conservation and Management Act; other regulations are published under the Marine Mammal Protection Act and Endangered Species Act. The NMFS also regulates fisheries pursuant to decisions of "regional fishery management organizations"(RFMOs) and other RFMOs to which the U.S. is a party, such as the Inter-American Tropical Tuna Commission, the International Commission for the Conservation of

Atlantic Tunas, the Western and Central Pacific Fisheries Commission, the Agreement on the International Dolphin Conservation Program, etc.

Some regulations by NMFS are quite controversial. For example, in 2007, the NMFS issued regulations to protect endangered whales from fatal fishing-gear entanglements after environmental groups sued to force action on the rules, which were proposed in early 2005. The rules were enacted to specifically protect the North Atlantic Right Whale, of which about only 350 remain. Marine gear entanglements and ship strikes are the top human causes of right whale deaths. On July 1, the shipping lanes in and out of Boston Harbor were rotated to avoid an area with a high concentration of the right whales.

The Central Marine Fisheries Research Institute

The Central Marine Fisheries Research Institute established by the Government of India under the Ministry of Agriculture in 1947 became a member of the Indian Council of Agricultural Research (ICAR) family in 1967. The headquarters was shifted from Mandapam Camp to Cochin in 1971.

Over the period, the Institute has grown significantly in its size, stature & research infrastructure. Now, it enjoys the status of a premier research organisation comparable to any similar institution in the developed countries.

With the fast changing marine fisheries scenario and emergence of mariculture as a viable proposition for enhancing the production, the Institute over the years has tuned its mandate in consonance with the needs of the time.

BANGLADESH MARINE FISHERIES ACADEMY

Bangladesh Marine Fisheries Academy(BMFA) is a government-run training institution in Bangladesh for cadets wishing to enter the fishing industry, merchant shipping and other related maritime industries. Established in 1973, it is the only national organization offering training for these professions. Fishing in Bangladesh is an important industry for a country with a shoreline on the Indian Ocean. Therefore, to explore and exploit the sea resources from 41,000 square miles (110,000 km2) of Exclusive Economic Zone, Bangladesh required to train personnel for our Fishery Sector which, inspired Government to establish Marine Fisheries Academy (MFA).

After the Bangladesh war of 1971 that established the country of Bangladesh, it was felt that the country lacked skilled manpower for the judicious exploitation and harvesting of marine fisheries resources. From 1972 there was a call for a national institution to train and educate personnel for Bangladesh maritime fisheries industries. The year was also the beginning of a rapid technological change in the deep sea fishing, with the introduction of large and sophisticated deep sea fishing vessels. The government realized that without qualified officers the growth of an efficient fishing fleet would be impossible. Therefore on 1 September 1973 the Marine Fisheries Academy was established in a beautiful setting on the southern bank of the Karnaphuli River, across from Sadarghat, Chittagong. The academy recruited its first batch of cadets in 1974.

The academy possesses extensive maritime and commercial fishing training facilities and prepares industry-based experts through teaching. Students are trained for careers as navigator, engineer or fish processing technologist for employment in Bangladesh or in other countries. Academy has built up an enviable reputation as the provider of excellent trained manpower.

Its geographical position is 22 19.0′ N latitude, 91 49.8′ E longitude, which is about 1 km from the city center.

Administrative control

Academy started its journey as a project of Bangladesh Fisheries Development Corporation (BFDC) and on 20 April 1988 it came under the administrative control of Ministry of Fisheries and Livestock as a project. But on 1 July 1993 the Academy was transferred to revenue budget under the same Ministry. Since then the Academy is functioning under the Administrative control of the Ministry. The functional authority of the Academy is vested on the Principal who has three assisting organs namely Academic Council, Board of Studies and Board of Discipline. The Academy has officers and staffs to carry out the instructional duties as well as administrative functions. To cover up the shortage of permanent instructors the Academy has number of required regular visiting lecturers as well. The Academy is having three departments and an administrative wing.

Curriculum

The academy offers a range of bachelor degree courses under Bangladesh National University, Gazipur. These are accredited courses, recognized throughout Bangladesh and elsewhere. Professional qualifications are recognized by the appropriate professional bodies throughout the world. These courses are for those who want to serve at sea on board fishing/ merchant vessels and on shore as processing technologists and so on.

The total duration of training of the cadets is three years. Out of that 1st and 2nd year training is conducted in the Academy to cover the theoretical aspects of professional and academic subjects. On completion of each year, they attend year final examination conducted by the National University. In the 3rd year they undergo practical sea training for 24 weeks on board deep sea fishing vessels and the rest of the 3rd year covers industrial training, practice and ancillary courses. On successful completion of these courses they attend final examination under National University. Thereafter the Marine Fisheries Academy training culminates with the passing out parade and award of graduation certificates.

Training in Marine Fisheries Academy has four dimensions. Through out their stay in the academy cadets are trained on four major aspects as shown below: a) Character Building : Development of sound and balanced character, truthfulness, honesty, hard work and dedication to duty and self respect has maximum focus in MFA training. (b) Professional Training : MFA syllabus includes the professional subjects like Maritime Science (Nautical), Ship Board Operations, Maritime Science (Engineering), Engineering Sciences, Marine Science and Post Harvest Technology. (c) Academic Training : MFA teaches the subjects of English, Physics, Mathematics, General Zoology, Bio-Chemistry and Bangladesh Studies. (d) Physical Training : MFA conducts physical training and parade, which contributes to the physical development, in addition it instills qualities like coolness, alertness, courage, self control, concentration, companionship and healthy sense of team spirit in the Cadets.

B.Sc.(Pass) Nautical

This is a comprehensive course for nautical education which provides academic and practical training on Navigation to the trainee cadet officers. Cadets of this course are trained up in such a way that they can navigate the ship by using stars and planets whenever satellite dependent electronic equipment betrays. This is a 3 (Three) years duration B.Sc.(Pass) course.

Cadets passed out from this Academy are serving in different capacity on boar Fishing and Merchant vessels both in home and abroad. At the beginning of the career on completion of requisite service cadets become eligible to appear in the Certificate of Competency (COC), conducted by Department of Shipping, Dhaka to achieve higher positions on board ships. Moreover experienced Navigators serve in various shore-based organization as Manager, Director and Adviser and as Instructor in Maritime Training Institutions.

B.Sc.(Pass) Marine Engineering

Cadets of this course complete a three-year full-time course in operation, maintenance and repair of marine engines and ship propulsion systems. Cadets undergo at least 24 weeks of industrial/sea practices. They also do sufficient workshop practices.

After successful completion of three years course at the Academy, the Engineering cadets can find jobs in diverse positions as Engineer on board Fishing or Merchant vessel. Besides finding employment on board ship, they will find a good opportunity in shore-based industries.

B.Sc.(Pass) Marine Fisheries

This is fully integrated professional course for the persons who are willing to prepare themselves for variety of careers in the fishing Vessels, Fish Processing Plants, Aquaculture Farm and Hatcheries etc. Cadets get sufficient practical training at sea and in various processing plants and aquaculture industry.

After successful completion of this course the passed out cadets can accommodate themselves into the commercial Fishing Industry, Fish Processing Industry, Hatchery, Aquaculture Farm, NGO's Processing Technologists, Quality Control Officers, Managers and Consultant.

Admission

Applications are invited from interested male citizens of Bangladesh for admission in B. Sc. (Pass) Nautical, B. Sc. (Pass) Marine Engineering and B. Sc. (Pass) Marine Fisheries Courses of Marine Fisheries Academy under the affiliation of National University , Gazipur.

Written Test will be held in two centers in Dhaka (Note Dame College) and Chittagong at the same date and time. One candidate may apply for any one course or for all three courses. This test is conducted on HSC standard subjects like English, Physics, Chemistry and General Knowledge which are common subjects for both Nautical/ Marine Engineering and Marine Fisheries Groups, Mathematics for Nautical/ Marine Engineering Group and Biology for Marine Fisheries Group. If someone opts for all three Departmental courses then he has to appear all six subjects.

For Nautical and Engineering Course: The Applicant must pass SSC and HSC examination with Mathematics minimum in 2nd Div or with GPA 2.5. In case any candidate had mathematics as forth subject he has to obtain at least 40% marks in that subject. For Marine Fisheries Course: The Applicant must pass SSC and HSC examination with Biology minimum in 2nd Div or with GPA 2.5. In case any candidate had Biology as forth subject he has to obtain at least 40% marks in that subject. Candidates appeared HSC Examination in 2006 and fulfill the above criteria may apply but he has to attach a certificate from the Head of his last Educational Institute that he has appeared HSC exam and expected to pass with minimum GPA 2.5.

Maximum age limit of the candidate is 20 years on the date as mentioned in the newspaper advertisement. Candidate should be of good health and having minimum height and weight 162.5 cm and 50 kg respectively.

Discipline

Ability to command the subordinates and carry out orders and superiors with utmost speed, promptness and judgment is of paramount importance for marine officers on board ship. This can only be expected from those who have developed complete adherence to the norms of discipline, which is essentially regarded as cumulative display of qualities of head and heart, sound training and professional competence. As such, due emphasis is laid on these aspects and the whole training programme and the activities of the Academy are designed to achieve this goal i. e. a high standard of discipline as in the Maritime Academies and similar other training institutions of the world. The Principal of the Academy reserves the right of dismissing any cadet at any time from the course or awarding any other suitable punishment in the event of breaking the rules of discipline of the Academy.

Cadets are granted shore leave (liberty) during weekend and other Government holidays from 0900 hrs to 1800 hrs. This facility allows the cadets to proceed ashore. A leave or liberty may be curtailed if a cadet's conduct is found unsatisfactory. Principal may extend, reduce or cancel leave or liberty when it seems that such action would be in the best interests of the Academy.

Graduation

On completion of the three-year course in the Academy the cadets are given the B. Sc. (Pass) Degree from National University, and are given Pre-Sea Nautical / Marine Engineering / Marine Fisheries certificates through a ceremony, which is called the "Passing-Out Parade." This is the most colorful event and a very special occasion for the cadets. During the passing-out parade all cadets of the Academy carry out different colorful march-pasts, like the cadets of the Armed Forces Academy, and show marks of respect to the Chief Guest of the Ceremony in the presence of distinguished guests and the cadets' parents. With an internationally accepted Pre-sea training certificate and a degree from the National University, graduates are able to find jobs both domestically and overseas. As of 2008, approximately 40% of academy's previous graduates are working in different organizations and capacities all over the world.

Medals and prizes

During passing out parade following medals / prizes are awarded to the deserving cadets: The Gold Medal for the best all round cadet of the

Academy. The best cadet is also awarded a crest from the Marine Fisheries Academy.

- The Silver Medal for the best cadet in the Nautical branch.
- The Silver Medal for the best cadet in the Engineering branch.
- The Silver Medal for the best cadet in the Fisheries branch.

Now top 10 cadets from Nautical and 10 Cadets from Engineering Branch is given the CDC by the academy. Rest of the passed out cadets will gradually get the CDC after serving one year in Fishing Vessels.

Now The cadets of BMFA also get CDC from Department of Shipping after their successful passing-out from the academy. So they are now serving on board merchant vessel too. Commander M Hassan Zaman Khan, (X), BN is the principal of the academy.

GULF STATES MARINE FISHERIES COMMISSION

The Gulf States Marine Fisheries Commission was established by an act of Congress (P.L. 81-66) in 1949 as a compact of the five Gulf States. Its charge is:

"to promote better utilization of the fisheries, marine, shell and anadromous, of the seaboard of the Gulf of Mexico, by the development of a joint program for the promotion and protection of such fisheries and the prevention of the physical waste of the fisheries from any cause."

The Commission is composed of three members from each of the five Gulf States. The head of the marine resource agency of each state, a member of the legislature, and a citizen with knowledge of marine fisheries appointed by the governor.

The Commission is empowered to make recommendations to the governors and legislatures of the five Gulf States regarding the management of the fisheries. The states do not relinquish any of their rights or responsibilities in regulating their own fisheries, however. Recommendations to the states are based on scientific studies made by experts employed by state and federal resource agencies and advice from law enforcement officials and the commercial and recreational fishing industries.

In addition, the Commission advises the United States Congress and may testify on legislation and marine policies that affect the Gulf. One of the most important functions of the Commission is to serve as a forum for the discussion of various problems, issues, and programs concerning marine management.

MARINE STEWARDSHIP COUNCIL

The Marine Stewardship Council (MSC) is an independent non-profit organization with an ecolabel and fishery certification programme. Fisheries that are assessed and meet the standard can use the MSC blue ecolabel. The MSC mission is to 'reward sustainable fishing practices'. When fish is bought that has the blue MSC ecolabel, it should indicate that this fishery operates in an environmentally responsible way and does not contribute to the global environmental problem of overfishing. As of the end of 2010, more than 1,300 fisheries and companies had achieved a Marine Stewardship Council certification.

Principles

The MSC environmental standard for sustainable fisheries was developed over two years through a consultative process. The three principles of the MSC standard consider:

1. The condition of the fish stock(s) of the fishery
2. The impact of the fishery on the marine ecosystem
3. The fishery management system

The MSC standard is consistent with the 'Guidelines for the Eco-labelling of Fish and Fishery Products from Marine Wild Capture Fisheries' adopted by the UN Food and Agriculture Organisation (FAO) in 2005. Any fishery that wishes to become MSC certified and use the ecolabel is assessed against the MSC standard by a third party, independent certification body that has been independently accredited to perform MSC assessments by Accreditation Services International (ASI). Chain of custody certification along the supply chain from boat to point of sale ensures that seafood sold bearing the ecolabel originated from an MSC certified fishery.

The MSC is headed by the Chief Executive who reports to the Board of Trustees. This is supported by the Technical Advisory Board and the Stakeholder Council, (further information on MSC governance). The MSC programme works through a multi-stakeholder partnership approach, taking into account the views of those seeking to secure a sustainable future.

History

The MSC was founded in 1997 by the World Wide Fund for Nature and Unilever, and became fully independent in 1999. It has offices in London (its head office), Seattle, Washington and Sydney, New South Wales and local offices in Edinburgh, Berlin, The Hague, Paris, Cape Town and Tokyo.

As of October 2010, there are 7,220 seafood products available with the MSC ecolabel, sold in 74 countries around the world. 97 fisheries have been independently certified as meeting the MSC's environmental standard for sustainable fishing and 132 are currently undergoing assessment, (further information on fisheries engaged in the MSC programme). 1,535 companies have met the MSC Chain of Custody standard for seafood traceability.

MSC and Aquaculture

The Marine Stewardship Council is a programme for wild fisheries and does not include aquaculture production. However, the MSC programme can contribute to sustainable outcomes along the aquaculture production chain through the certification of feed fish fisheries. As of October 2010 there is one fishery undergoing assessment for MSC certification that produces for fishmeal - the Gulf of California sardine fishery. About 85% of the total production from the Gulf of California sardine fishery is used for reduction to fishmeal and mostly used for animal feeds.

Fish & Kids

Fish & Kids is a project of the Marine Stewardship Council that encourages schools and restaurants in England and Sweden to serve sustainable, MSC labelled seafood. Fish & Kids also aims to raise children's awareness of sustainable food issues through a classroom resource pack and website using fun games and educational materials.

Independent Opinion and Criticism

Jared Diamond's Collapse praises MSC and the similar Forest Stewardship Council as good examples of collaboration among environmentalists and businesses for a sustainable economy. In early 2010, the MSC was criticised for including the British Columbia sockeye fishery in the certification process. This is because in 2009, just a fraction of the salmon returned to the Fraser River, which prompted the Prime Minister to launch a judicial enquiry.

Scientists at a Simon Fraser University (SFU) summit of Fraser River Sockeye Salmon said of the low salmon run: "In 2009 fisheries responded appropriately by greatly restricting fishing to maximize the number of fish available for spawning. The weight of evidence suggests that the problem of reduced productivity occurred after the juvenile fish began their migration toward the sea."

The total 2010 Fraser River sockeye return was over 35 million fish; the largest return since 1913. MSC certification of the B.C. sockeye fishery is confirmation that the management of the fishery aims to ensure that fish stocks remain above critical lower limit levels. Certification was not a conclusion that the stock was currently abundant or that fishing should be taking place at that time.

Friend of the Sea

Friend of the Sea is a project for the certification and promotion of seafood from sustainable fisheries and sustainable aquaculture. It is the only certification scheme which, with the same logo, certifies both wild and farmed seafood. Friend of the Sea started as a project of the Earth Island Institute, the NGO which operates the successful International Dolphin-Safe project. Some of the main world retailers participate, such as Carrefour Italy, Coop Italia, Eroski, Manor and Finiper. Some important producers also have their products certified.

Compliance and criteria

Friend of the Sea's criteria compliance is verified by independent accredited certification bodies. Essential criteria for fisheries are:

- The product should not originate from overexploited (nor depleted, data deficient or recovering) stocks
- The fishing method should not impact the seabed
- The fishing method should be selective (below the world average for discards, which in 2005 was about 8%).
- The fishery should respect all legal requirements.

Article 30 of the FAO Guidelines for Ecolabelling of Marine Fisheries Products requires that the target stock in fisheries is not overexploited. The Friend of the Sea sustainable seafood certification scheme is the only certification scheme currently compliant with this article.

SEACHOICE

SeaChoice is a program of Sustainable Seafood Canada that produces sustainable seafood advisory lists. The program is designed to raise consumer awareness about the importance of buying seafood from sustainable sources. It is best known for publishing consumer guides for responsible seafood purchasing. SeaChoice evaluates fisheries based on habitat/ecosystem impacts, stock status, inherent vulnerability, bycatch, and management regimes.

Sustainable Seafood Canada is a Canadian coalition of environmental organizations working together to raise awareness and deliver solutions for sustainable fisheries. The organization consists of the Canadian Parks and Wilderness Society, the David Suzuki Foundation, the Ecological Action Centre, Living Oceans Society and British Columbia's chapter of the Sierra Club of Canada. SeaChoice collaborates with the Monterey Bay Aquarium's Seafood Watch program.

Seafood list

The group gives Canada-centric lists of recommendations for the best seafood choices, fish which have "some concerns" and fish to avoid. The "avoid" category is reserved for seafood which is overfished and/or fished or farmed in ways that harm other marine life or the environment. Health alerts for fish with high levels of contaminants (e.g. mercury, dioxins, PCBs) are also noted, although they may appear in any category. The SeaChoice website includes both regional and country-wide guides for Canada. The organization also produces a guide for sushi. Pocket guides are also available from the organization.

Recommended seafood from SeaChoice includes Alaska pollock, longline Pacific cod, land-based rainbow trout, troll or pole-caught albacore tuna, United States spiny lobster, farmed Arctic char, Pacific herring and sardines.

Partnerships

In March 2010, SeaChoice announced an agreement with Federated Co-operatives, an organization representing over 250 co-operatives. The agreement includes a sustainability plan for seafood sales and procurement. SeaChoice has also partnered with Overwaitea Food Group, a supermarket chain in Western Canada. Bento Nouveau, Canada's largest sushi company, partnered with SeaChoice in June 2010.

SEAFOOD WATCH

Seafood Watch is one of the best known sustainable seafood advisory lists, and has influenced similar programs around the world. It is a program designed to raise consumer awareness about the importance of buying seafood from sustainable sources. It is best known for publishing consumer guides for responsible seafood purchasing in the United States, including making them available on mobile devices, such as the iPhone.

Seafood Watch is a program of the Monterey Bay Aquarium, and is a partner of SeaWeb's Seafood Choices Alliance. It has roots in the Monterey Bay Aquarium's Fishing for Solutions exhibit which ran from 1997 to 1999 and produced a list of sustainable seafood. It was one of the first resources for sustainable seafood information together with the Audubon Society's What is a fish lover to eat? which also came out in the late 1990s.

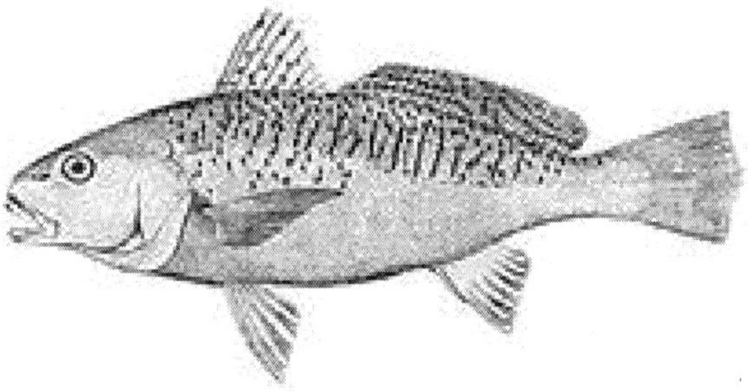

Fig—The fast-growing and resilient Atlantic croaker

Seafood List

The group gives somewhat US-centric lists of recommendations – the best seafood choices, fish to avoid, as well as "good alternatives". The "avoid" category is for seafood which is overfished and/or fished or farmed in ways that harm other marine life or the environment. Health alerts for fish with high levels of contaminants (e.g. mercury, dioxins, PCBs) are also noted, although they may appear in any category.

The Seafood Watch website includes both regional and country-wide guides for the United States. Pocket guides are available from the aquarium and further information is on the web site. Several of the regional guides are also available in Spanish. The guides are updated twice annually, while the website is updated more often. Recommended seafood includes Sardines, US-farmed Sturgeon (but not wild caught), Atlantic Croaker, Pacific Halibut, Wreckfish, White Seabass and Dungeness Crab. Restaurants and retailers are also targeted with an educational program developed by Seafood Watch.

In 2010 Seafood Watch added its "Super Green" list, which features seafood that it is good for human health and does not harm the oceans. The Super Green list highlights products that are currently on the Seafood Watch "Best Choices" (green) list, are low in environmental contaminants and are good sources of long-chain omega-3 fatty acids.

Industry Criticism

Recently, industry organizations have begun to push back against Seafood Watch's efforts. After publication of a sustainable sushi guide, the National Fisheries Institute, a seafood industry trade group, wrote on its blog that the guides were "confusing and contradictory," adding that they didn't fully take into account the economic, environmental and social aspects of seafood sustainability.

In December 2008, the British Columbia Salmon Farmers Association published a letter from executive director, Mary Ellen Walling to Geoff

Shester, the former Senior Science Manager of Monterey Bay Aquarium's sustainable seafood initiative. Walling outlined several concerns about Seafood Watch's methods, specifically questioning the red list designation of farmed Atlantic salmon when British Columbia-based farms shared so many of the same methods with those used in Alaska to produce ranched salmon. Walling also contended that Monterey Bay's Report on farmed Atlantic salmon was woefully out of date and that Monterey Bay needed to demonstrate greater transparency in its review process in order to continue to be considered credible by industry.

OCEANA

Oceana is the largest international ocean conservation and advocacy organization. Oceana works to protect and restore the world's oceans through targeted policy campaigns.

Oceana bases its policy campaign goals on science to achieve concrete and measurable results through targeted campaigns that combine policy advocacy, science, law, media, and public pressure to prevent collapse of fish populations, marine mammals, and other sea life caused by industrial fishing and pollution. Campaigns are designed to produce clear, identifiable policy changes within a 3-5 year timeframe.

Oceana is headquartered in Washington, D.C. and has North American offices in New York, NY, Juneau, AK, Anchorage, AK, Portland, OR, Monterey, CA, Boston, MA and Los Angeles, CA. In Europe, Oceana has offices in Brussels, Belgium and Madrid, Spain. The South American office is in Santiago, Chile and the Central American office is in Belize City, Belize.

Oceana was established in 2001 by a group of leading foundations-The Pew Charitable Trusts, Oak Foundation, Marisla Foundation (formerly Homeland Foundation), and the Turner Foundation. Those foundations had discovered through a study commissioned in 1999 that less than one-half of one percent of all resources spent by environmental non-profit groups in the United States went to ocean advocacy. Thus Oceana was created to identify practical solutions to the problems facing the oceans and to make those solutions happen.

The organization was not started from scratch, as the Ocean Law Project-also initiated by The Pew Charitable Trusts-was absorbed into Oceana in 2001 as the Oceana's legal arm. In 2002, Oceana merged with American Oceans Campaign, founded by actor/environmentalist Ted Danson, to more effectively address our common mission of protecting and restoring the world's oceans. Danson remains a committed and active member of Oceana's Board of Directors.

Campaigns

- The Ocean Pollution campaign is working to stop major sources of ocean pollution including oil, mercury, aquaculture and shipping emissions.
- The Responsible Fishing campaign is working to promote sustainable fishing practices, which includes stopping destructive bottom trawling and illegal fishing gear, limiting bycatch, ending fishing subsidies and protecting overfished species such as bluefin tuna and krill.
- The Sharks campaign works to protect sharks through policy, science and legal work. Oceana is pushing for true shark finning bans, species-specific shark management and reduced shark bycatch.
- The Sea Turtles campaign works to protect sea turtles from extinction by reducing sea turtle bycatch, protecting habitat and develop legislation to protect sea turtles.
- The Climate and Energy campaign works to stop ocean acidification, promote clean alternative energy such as offshore wind, and end offshore drilling.
- Oceana also campaigns to protect vulnerable places in the world's oceans, including the Arctic, the Aleutian Islands, Chile's Juan Fernandez Islands, Punta de Choros and Patagonia, and the Mediterranean.
- Every year since 2005, the organization has led an on-the-water expedition in the Mediterranean to document vulnerable marine habitats. In 2010, Oceana spent two months on the water in the Gulf of Mexico investigating the long-term effects of the Deepwater Horizon oil spill.

Past Campaign Work and Victories

North America

- Protecting Sea Turtles: In March 2010, in response to two petitions submitted in 2007 by Oceana and others, the National Marine Fisheries Service and the Fish and Wildlife Service issued a proposed rule to change the status of North Pacific and Northwest Atlantic loggerhead sea turtles from "threatened" to "endangered" under the Endangered Species Act.
- Freezing the Footprint in the Arctic: In 2008, the National Marine Fisheries Service closed nearly 115 million acres (470,000 km2) of the Bering Sea to destructive bottom trawling. The closure adopted Oceana's approach of freezing the current "footprint" already trawled and prevents future expansion of this damaging fishing technique. Worldwide, Oceana has protected more than 640 million acres (2,600,000 km2) of seafloor from bottom trawling.

In addition, after work by Oceana, in February 2009, the North Pacific Fishery Management Council voted to prevent the expansion of industrial fishing into all U.S. waters north of the Bering Strait for the foreseeable future to limit stress on ocean ecosystems in light of the dramatic impacts of global climate change in the Arctic.

- Getting the Word out About Mercury: Oceana has convinced major grocery retailers including Costco, Kroger, Whole Foods, Trader Joe's, Safeway, Wild Oats, Harris Teeter and others to post the Food and Drug Administration's warning regarding mercury in certain kinds of fish at their seafood counters. As a part of this campaign, Oceana also targeted outdated chlorine manufacturing plants that the organization believes to be polluting the environment with mercury.
- Oceana is also known for their involvement in campaigning Royal Caribbean to engage in environmentally-friendly waste management. After a year of campaigning by Oceana, Royal Caribbean agreed to install advanced wastewater treatment facilities on all of its ships.

South America

- Preserving Punta de Choros: In August 2010, Oceana helped convince Chilean President Sebastiàn Piñera to request the relocation of a coal-fired power plant scheduled to be constructed in a vulnerable region of Chile's northern coast, Punta de Choros.
- Regulating Salmon Aquaculture: In March 2010, as a direct result of Oceana's campaign work to reform the Chilean salmon aquaculture industry, the Chilean Congress passed legislation to prevent the escape of farmed salmon and further regulate the use of antibiotics in salmon aquaculture.

Europe

- Saving Sharks: In January 2009, the Spanish government, after campaigning and consulting with Oceana, committed to advancing new shark legislation that would ban the catch of threatened hammerhead and thresher sharks, put in place catch limits for blue sharks and shortfin mako sharks and evaluate the viability of landing sharks "whole" with their fins attached. Spain is one of the largest shark catching and exporting countries in the world.
- Banning Mediterranean Driftnetting: In February 2008, the European Court of Justice rejected any further requests by the French government for exemptions from the EU ban on driftnetting in the Mediterranean Sea. This ruling will spare 25,000 juvenile bluefin tuna annually, along with 10,000 non-targeted marine species caught annually in the driftnets.

- Limiting Destructive Trawling: After two years of intensive lobbying by Oceana staff in Brussels and Madrid, in 2005 the European Union prohibited destructive fishing practices, including bottom trawling, in over 500,000 square miles (1,300,000 km2) around the Azores, Madeira and the Canary Islands.
- Preventing Oil Pollution: In May 2005, a new law was passed by the European Union that imposes criminal sanctions on the owners, operators and financiers of boats that illegally dump oily waters and residues into the sea. The law could prevent as much as 20,000,000 tons of polluting substances from getting into the ocean every year.

Notable Supporters

Actors Ted Danson and Sam Waterston are both on Oceana's board of directors, but a variety of notable celebrities are supporters of the organization. "Mad Men" actress January Jones is the face of Oceana's "Scared for Sharks" campaign, and "Private Practice" star Kate Walsh filmed a PSA about sea turtles for Oceana. In 2010, "Entourage" actor Adrian Grenier swam with bluefin tuna and filmed a PSA about the endangered fish.

In addition, U.S. Swimmer Aaron Peirsol has led an open-water swim event for Oceana since 2007 called "The Race For the Oceans." Other Oceana supporters include Ben Harper, Morgan Freeman, and "End of the Line" author Charles Clover.

WORLDFISH CENTER

The WorldFish Center is a non-profit international research organization that collaborates with partners to carry out scientific research on fisheries and aquaculture aimed at reducing poverty and hunger in developing countries. The Center delivers science-based services and solutions (knowledge, expertise, methods, tools, technologies) directed towards solving poverty, hunger and environmental degradation in Africa, Asia and the South Pacific.

WorldFish is one of 15 research centers of the Consultative Group on International Agricultural Research (CGIAR) funded by donor countries, multilateral and global organizations and foundations.

The Approach

Working in partnership with private and public sectors and civil society, WorldFish develops pro-poor sustainable aquaculture that supports the Millennium Development Goals (MDGs). All services and solutions developed by the Center are international public goods that are made freely available to all.

WorldFish takes a comprehensive, multidisciplinary research approach that acknowledges the complex and multi-faceted problems that face fisheries and aquaculture. Failure to embrace this complexity has led to piecemeal efforts in the past and undue faith in single technology or development approaches. The Center prioritizes its research efforts to include those areas in which it will have the biggest impacts, and assumes the role of broker and catalyst of research among the full range of development partners needed to close the gap between research and development action.

Why Fish?

The world's bottom billion are falling behind in the war on poverty and hunger. Traditional staple foods alone cannot feed the world while fisheries and aquaculture have tremendous potential to provide the poor with more food, better nutrition and increased incomes.

Fisheries are a source of high-protein food: they provide more than 1.5 billion people with almost 20 percent of their average per capita intake of animal protein, and 3 billion people with at least 15 percent of such protein. In some countries they provide up to 70% of animal protein -affordable, high-quality animal protein- that is particularly important for mothers-to-be and young children.

Catching, processing and trading fish provide a source of income and livelihood for millions of men and women, the overwhelming majority of whom are associated with small-scale fisheries. In fact, of the 70% of the world's total fish catch that comes from developing countries, over a half of this comes from small-scale fisheries.

The Challenges

Small scale fisheries have failed to keep pace with demand for food and employment in developing countries and urgently require attention to technology, health, economics and reform in management and governance. Over 20% of the world's 38 million full-time fishers are among the world's poorest people, earning less than US$1 per day, but most wild-fish stocks are reaching the limits of their productive capacity even as many have stabilized or declined because of overfishing and other causes. While the capacity needed to maintain wild fisheries must be amplified in the face of these constraints and pressures, development of aquaculture is also necessary as it provides the only option for substantial increases in fish production.

Climate change poses additional challenges that will affect the productivity, distribution and seasonality of fisheries and the quality and availability of the habitats that support them while a further global threat comes from degradation of coral reefs. Seventy-five percent of coral reefs are currently under immediate and direct threat from local sources, which include overfishing, destructive fishing, coastal development, and pollution.

WorldFish Research

WorldFish expertise in Policy Economics and Social Sciences, Natural Resource Management and Aquaculture and Genetic Improvement work together to provide a wide range of research and analysis to meet complex challenges like these. The Policy Economics and Social Science division focuses on social and economic analysis of the aquaculture and fisheries sectors; connecting the fisheries and aquaculture sector to poverty reduction initiatives at local to global scales; policy and institutional analysis for the improved governance of aquatic resources; assessing the potential impacts of climate change on fisheries and adaptive measures that can be taken; and human health consequences of fisheries, reducing risks, and fisheries options that benefit health-impaired populations (HIV/AIDS and malaria). WorldFish also works with communities to manage their fisheries.

The Natural Resources division produces integrated assessment and management technologies for small-scale fisheries, designs and manages global information systems on aquatic resources (FishBase , ReefBase), studies post-disaster livelihood recovery in fisheries-dependent regions, assesses impacts of built structures on aquatic resources in river basins and analyses external drivers such as climate change on livelihoods of fishery-dependent households. The Aquaculture and Genetic Improvement division is dedicated to developing methods for breeding improved fish strains for aquaculture; aquaculture technologies for the poor, including women and the landless; integrating aquaculture with terrestrial small-scale agriculture; developing strategies and options for aquaculture production and national action plans; connecting small-scale producers to markets and developing technologies that improve water productivity while protecting environmental flows.

Impact and innovation

WorldFish, with its partners, has raised incomes for millions of poor people (and reduced suffering of HIV/AIDS-affected families) by integrating aquaculture with agriculture and has empowered poor communities to participate in the sustainable co-management of their fisheries. It has helped countries cope with disaster and conflict by restoring fisheries, provided nations with tools to improve the planning and management of major river basins and developed widely-consulted global databases and strengthened national capacities for fisheries management.

Three areas of work have generated particularly large impact: The breeding of much higher-yielding tilapia fish varieties (GIFT), widely used in aquaculture across Asia, greatly raising productivity and incomes: $170 returned for each $100 invested per annum. Integrated aquaculture-agriculture in Malawi that has sharply increased incomes and reduced childhood malnutrition, and helping HIV/AIDS-affected families cope; $115 returned for each $100 invested per annum. Fisheries co-management in Bangladesh, which is increasing biodiversity, raising incomes by 100% and fish catches by 30%, particularly by empowering women. The Science Council commended co-management as an "eminently replicable model for contemporary rural development."

Defying Ocean's End

Defying Ocean's End (DOE) is a global agenda for action in marine conservation compiled in a 2004 Island Press book. It is also the title of a 2003 Los Cabos (Mexico) conference, where the agenda was formulated. The DOE website describes itself as "a practical agenda of action to safeguard the ocean for future generations to come. This dynamic strategy-including costs and recommendations-is a collective voice for those seeking to reverse the disturbing trends we are witnessing in the ocean today. As never again, we have an opportunity now to respond to this crisis, by moving beyond localized and ad hoc initiatives - however good they might be - to coordinated global action." The website also lists a "DOE Office" charged with distributing the agenda. The office is composed of Dr. Sylvia Earle, Arlo Hemphill, Tim Noviello and Linda K. Glover: .

HOTSPOT ECOSYSTEM RESEARCH AND MAN'S IMPACT ON EUROPEAN SEAS

Hotspot Ecosystem Research and Man's Impact On European Seas, or HERMIONE, is an international multidisciplinary project, started in April 2009, that studies deep-sea ecosystems. HERMIONE scientists study the distribution of hotspot ecosystems, how they function and how they interconnect, partially in the context of how these ecosystems are being affected by climate change and impacted by humans through fishing, resource extraction, seabed installations and pollution. Major aims of the project are to understand how humans are affecting the deep-sea environment and to provide policy makers with accurate scientific information, enabling effective management strategies to protect deep sea ecosystems.The HERMIONE project is funded by the European Commission's Seventh Framework Programme, and is the successor to the HERMES project, which concluded in March 2009.

Europe's deep-ocean margin, from the Arctic to the Iberian Margin, and across the Mediterranean to the Black Sea, spans a distance of over 15,000 km and hosts a number of diverse habitats and ecosystems. Deep water coral reefs, undersea mountains populated by a multitude of organisms, vast submarine canyon systems, and hydrothermal vents are some of the features contained therein. The traditional view of the deep-sea realm as a hostile and barren place was discredited long ago, and scientists now know that much of Europe's deep sea is rich and diverse.

However, the deep sea is increasingly threatened by humans: most of this deep-ocean frontier lies within Europe's Exclusive Economic Zone (EEZ) and has significant potential for the exploitation of biological, energy, and mineral resources. Research and exploration over the last two decades has shown clear signs of direct and indirect anthropogenic impacts in the deep sea, resulting from such activities as over-fishing, littering and pollution. This raises concerns because deep-sea processes and ecosystems are not only important for the marine web of life, but also fundamentally contribute to the global biogeochemical cycle.

Continuing with the knowledge obtained by the HERMES project (EC FP6), which contributed significantly to our understanding of deep-sea ecosystems, the HERMIONE project investigates ecosystems at critical sites on Europe's deep-ocean margin, aiming to make major advances in knowledge of their distribution and functioning, and their contribution to ecosystem goods and services. HERMIONE places special emphasis on human impact on the deep sea and on the translation of scientific information into science policy for the sustainable use of marine resources. To design and implement effective governance strategies and management plans to protect our deep seas for the future, understanding the extent, natural dynamics and interconnection of ocean ecosystems, and integrating socio-economic research with natural science, are important. To achieve this, HERMIONE uses a highly interdisciplinary and integrated approach, engaging experts in biology, ecology, biodiversity, oceanography, geology, sedimentology, geophysics and biogeochemistry, who will work alongside socio-economists and policy-makers.

Hotspot Research

The HERMIONE project focuses on deep-sea "hotspot" ecosystems including submarine canyons, open slopes and deep basins, chemosynthetic environments, deep water coral reefs, and seamounts. Hotspot ecosystems support high species diversity, numbers of individuals, or both, and are therefore important in maintaining margin-wide biodiversity and abundance. HERMIONE research ranges from investigation of the ecosystems' dimensions, distribution, interconnection and functioning, to understanding the potential impacts of climate change and anthropogenic disturbance. The ultimate objective is to provide stakeholders and policymakers with the scientific knowledge necessary to support deep-sea governance, sustainable management and conservation of these ecosystems.

To obtain the data needed, HERMIONE scientists are spending over 1000 days at sea, using more than 50 research vessels across Europe. Sharing vessels and equipment between partners will bring benefits through shared knowledge, expertise and data, and will also maximise the research effort, increasing efficiency and productivity. State-of-the-art technology will be used, with Remotely Operated Vehicles (ROVs) one of the critical pieces of equipment being used for a wide range of delicate manoeuvres and high-resolution surveys, from precision sampling of methane gas at cold seeps to microbathymetry mapping to examine the structure of the seabed. Large arrays of instrumented moorings, shared by different partner institutions, will be deployed in common experimental areas, allowing HERMIONE to develop experimental strategies beyond any national capacity.

Study Areas

The HERMIONE study sites were selected on the following basis:

- The Arctic because of its importance in monitoring climate change;
- Nordic margin with abundant cold-water corals, extensive hydrocarbon exploration and the Haakon-Mosby mud volcano (HMMV) natural laboratory;
- Celtic margin with a mid-latitude canyon, cold water corals and the long term Porcupine Abyssal Plain (PAP) monitoring site;
- Portuguese margin with the highly diverse Nazare and Setubal canyons;
- Seamounts in the Atlantic and western Mediterranean as important biodiversity hotspots potentially under threat;
- Mid Atlantic Ridge (MAR) ESONET site to link cold seep to hot seep chemosynthetic studies;
- Mediterranean cold water cascading sites in the Gulf of Lions and outflows of the Adriatic and Aegean Seas.

The HMMV, PAP, MAR and central Mediterranean sites link to the ESONET long-term monitoring sites and will provide valuable background information.

Hotspot Ecosystems

Cold-water coral reefs

Deep water coral reefs are found along the northeast Atlantic and central Mediterranean margins, and are important biodiversity hotspots. The recent HERMES project lists more than 2000 species associated with cold-water coral reefs worldwide. As well as flourishing live coral, the dead coral frameworks and rubble that are frequently found close by attract a myriad of fauna from the microscopic to the mega, and may be fundamental in coral ecosystem replenishment. Coral reefs provide a habitat for fish, a refuge from predators, a rich food source, a nursery for young fish, and are also potential sources of a wide range of medicines to treat ailments from cancer to cardiovascular disease.

There are several known coral hotspot areas on Europe's deep-ocean margin, including the Scandinavian, Rockall-Porcupine and central Mediterranean margins, and there remain many questions about them, such as how each of the sites are connected to one another, how they arose, what drives the distribution of the reefs, how the larvae disperse and settle, how the corals and associated species reproduce, finding their physiological thresholds, how they will fare with increased ocean warming, and whether ocean warming induces a spread of coral reefs further north into the Arctic Ocean. New research will also build on previous work to define the physical environment around cold-water coral reefs such as hydrodynamic and sedimentary regimes, which will help to understand biological responses.

HERMIONE scientists are using cutting-edge technology to try to answer these questions. High-resolution mapping of the seafloor will be carried out to determine the location and distribution of cold-water corals, and photographic observations will be made to assess changes in the status of known reefs over time, such as their response to climatic variation or their recovery from destruction by fishing trawlers. To assess biodiversity and its relationship with environmental factors such as climate change, DNA barcoding and other molecular techniques will be used.

Submarine Canyons

Submarine canyons are deep, steep-sided valleys that form on continental margins. Stretching from the shelf to the deep sea, they dissect much of the European margin. They are one of the most complex seascapes known to man; their rugged topography and challenging environmental conditions mean that they are also one of the least explored. Advances in technology over the last two decades have allowed scientists to uncover some of the mysteries of canyons, the size of which often rival the Grand Canyon, USA.

One of the most important discoveries is that canyons are major sources and sinks for sediment and organic matter on continental margins. They act as fast-track pathways for sediment and organic matter from the shelf to the deep sea, and can act as temporary depots for sediment and carbon storage. Particle flux through canyons has been found to be between two to four times greater than on the open slope, though the transfer of particles through canyons is thought to be largely "event-driven", which introduces a highly variable aspect to canyon conditions. Determining what drives sediment transport and deposition within canyons is one of the major challenges for HERMIONE.

The capacity of canyons to focus and concentrate organic matter can promote high abundances and diversity of fauna. However, variability in environmental conditions and topography is very high, both within and between canyons, and this is reflected in the variability of the structure and dynamics of the biological communities. Our understanding of biological processes in canyons has greatly improved with the use of submersibles and ROVs, but this research has also revealed that the relationships between fauna and canyons are more complex than previously thought. The diversity of submarine canyons and their fauna means that it is difficult to make generalisations that can be used to create policies for canyon ecosystem management. It is important that the role of canyons in maintaining biodiversity, and how potential anthropogenic impacts may affect this, is better understood. HERMIONE will address this challenge by examining canyon ecosystems from different biogeochemical provinces and topographic settings, in light of the complex interactions among habitat (topography, water masses, currents), mass and energy transfer, and biological communities.

Open slopes and deep basins

Open slopes and deep basins make up > 90 % of the ocean floor and 65 % of the Earth's surface, and many of the goods and services provided by the deep sea (e.g., oil, gas, climate regulation and food) are produced and stored by them. They are intricately involved in global biogeochemical and ecological processes, and so are essential for the functioning of our biosphere and human wellbeing.

Recent research in the HERMES (EC-FP6) project gathered a large body of information on local biodiversity at large scales, different latitudes and in different hotspot ecosystems, but the research also highlighted the high degree of complexity of deep-sea habitats. This information is fundamental to our understanding of the factors that control biodiversity at much larger scales, from hundreds to thousands of kilometres. HERMIONE will conduct further studies on the mosaic of habitats found in deep-sea slopes and basins, and will investigate the relationships within and between these habitats, their biodiversity and ecology, and their interconnection with other hotspot ecosystems.

Investigating the impacts of anthropogenic activities and climate change in the deep sea is a theme that runs through all HERMIONE research. To the biological communities on open slopes and in deep basins, seafloor warming through climate change is a major threat. Up to 85% of methane reservoirs along the continental margin could be destabilised, which would not only release climate-warming methane gas into the atmosphere, but would also have unknown and potentially devastating consequences on benthic communities. The role of climatic variation on deep-sea benthos is not well understood, although large-scale changes in the structure of seafloor communities have been observed over the last two decades. The use of long-term, deep-sea observatories, e.g., the Hausgarten deep-sea observatory in the Arctic and the time-series analysis of the Catalan margin and Southern Adriatic Sea, will help HERMIONE scientists to examine recent changes in benthic communities, and to study decadal variability in physical processes, such as the dense shelf water cascading events in submarine canyons.

HERMIONE aims to provide quantitative estimates of the potential consequences of biodiversity loss on ecosystem functioning, to examine how deep-sea benthos adapt to large-scale changes, and, for the first time, to create conceptual models integrating deep-sea biodiversity and quantitative analyses of ecosystem functioning and processes.

Seamounts

Seamounts are underwater mountains that rise from the depths of the ocean, and whose summits can sometimes be found just a few hundred metres below the sea surface. To be classified as a seamount the summit must be 1000 m higher than the surrounding seafloor, and under this definition there are an estimated 1000-2800 seamounts in the Atlantic Ocean and around 60 in the Mediterranean Sea.

Seamounts enhance water flow through localised tides, eddies, and upwelling, and these physical processes may enhance primary production. Seamounts may therefore be considered as hotspots of marine life; fauna benefit from the enhanced hydrodynamics and phytoplankton supply, and thrive on the slopes and summits. Suspension feeders, such as gorgonian sea fans and the cold-water corals like Lophelia pertusa, often dominate the rich benthic (seafloor-dwelling) communities. The enhanced abundance and diversity of fauna is not limited to benthic species, as fish are known to aggregate over seamounts. Unfortunately, this knowledge has led to increasing commercial exploitation of seamount fish by the fishing industry, and a number of seamount fish populations have already been depleted. Part of HERMIONE research will assess the threats and impacts of human activities on seamounts, including comparing data from seamounts in different stages of fisheries exploitation to understand more about the impacts of fishing activities., both on target species and non-target species, and their habitats.

Despite our increasing knowledge on seamounts, there is still very little known about the relationships between their ecosystem functioning and biodiversity, and that of the surrounding areas. This information is vital in order to improve our understanding of connectivity between seamount hotspots and adjacent areas, and HERMIONE research will aim to discover whether seamounts act as centres of speciation (the evolution of new species), or if they play a role as "stepping stones", allowing fauna to colonise and disperse across the oceans.

Chemosynthetic ecosystems

Chemosynthetic environments - such as hot vents, cold seeps, mud volcanoes and sulphidic brine pools - show the highest biomass and productivity of all deep-sea ecosystems. The chemicals found in the fluids, gases and mud that escape from such systems provide an energy source for chemosynthetic bacteria and archaea, which are the primary producers in these systems. A huge variety of fauna profits from the association with chemosynthetic microbes, supporting large communities that can exist independently of sunlight. Some of these environments, such as methane (cold) seeps, can support up to 50,000 times more biomass than communities that rely on photosynthetic production alone. Owing to the extreme gradients and diversity in physical and chemical factors, hydrothermal vents also remain incredibly fascinating ecosystems. HERMIONE researchers aim to illustrate the tight coupling between geosphere and biosphere processes, as well as their immense heterogeneity and interconnectivity, by observing and comparing the spatial and temporal variation of chemosynthetic environments in European Sea's.

Methane cycling and carbonate formation by microorganisms in chemosynthetic environments have implications for the control of greenhouse gases. Methane can be trapped and stored under the seabed as a gas hydrate, and under different conditions, can either be controlled by microbial consumption, or can escape into the surrounding seawater, and ultimately the atmosphere. Our understanding of the biological controls of methane seepage and feedback mechanisms for global warming is limited. The distribution and structure of cold seep communities can act as an indicator for changes in methane fluxes in the deep sea, e.g. by seafloor warming. Using multibeam echosounder data and 3D seismic data with in situ studies at seep sites, and by investigating the life histories of fauna at such ecosystems, HERMIONE scientists aim to understand more about their interconnectivity and resilience, and the implications for climate change.

The great variety of fauna present in chemosynthetic environments is a real challenge to scientists. Only a tiny fraction of microorganisms at vents and seeps has been identified, and a huge amount is still to be discovered. Their identification, their association with fauna, and the relationship between their diversity, function and habitat, are vital areas of research as biological communities act as important filters, controlling up to 100 % of vent and seep emissions. By using DNA barcoding and genome analysis in addition to traditional methods of identification and experimentation, HERMIONE scientists will study the relationship between community structure and ecosystem functioning at a variety of vents, seeps, brine pools and mud volcanoes.

Socio-economics, governance and science-policy interfaces

With increasing ocean exploration over the last two decades has come the realisation that humans have had an extensive impact on the world's oceans, not just close to our shores, but also reaching down into the deep sea. From destructive fishing practices and exploitation of mineral resources to pollution and litter, evidence of human impact can be found in virtually all deep-sea ecosystems. In response, the international community has set a series of ambitious goals aimed at protecting the marine environment and its resources for future generations. Three of these initiatives, decided on by world leaders during the 2002 World Summit on Sustainable Development (Johannesburg), are to achieve a significant reduction in biodiversity loss by 2010, to introduce an ecosystems approach to marine resource assessment and management by 2010, and to designate a network of marine protected areas by 2012. A crucial requirement for implementing these is the availability of high-quality scientific data and knowledge, as well as effective science-policy interfaces to ensure the policy relevance of research and to enable the rapid translation of scientific information into science policy.

HERMIONE aims to provide this by filling the knowledge gap about threatened deep-sea ecosystems and their current status with respect to anthropogenic impacts (e.g. litter, chemical contamination). Socio-economists and natural scientists work together in HERMIONE, researching the socio-economics of anthropogenic impacts, mapping human activities that affect the deep sea, assessing the potential for valuing deep-sea ecosystem goods and services, studying governance options and designing and implementing real-time science-policy interfaces.

HERMIONE natural and social science results will provide national, regional (EU), and global policy-makers and other stakeholders with the information needed to establish policies to ensure the sustainable use of the deep ocean and conservation of deep-sea ecosystems.

PROFISH

PROFISH is a global program on sustainable fisheries established by the World Bank in 2005. It has been set up in conjunction with key donors and stakeholders to meet the challenge of a growing crisis in the world fisheries sector.

The World Bank characterises this crisis in the following way:

"Increasing population pressures, growing demand for fish and failures of governance are leading to unsustainable levels of exploitation of living aquatic resources and destruction of aquatic ecosystems. In many developing countries the sustainable benefits are in decline, perpetuating a spiral into poverty for many small-scale fishers and communities dependent on fishing."

PROFISH is a programming and funding partnership between key fishery sector donors, international financial institutions, developing countries, stakeholder organizations, and international agencies. PROFISH currently receives financial and in-kind support from Iceland, France, Norway, Finland, Japan, FAO, World Bank, International Union for Conservation of Nature (IUCN) and WorldFish Center.

The overall objective of PROFISH is "to improve sustainable livelihoods in the fisheries sector and to make concrete progress towards meeting the goals of the WSSD's goals in fisheries".

The specific objectives are to strengthen governance of the world's marine fisheries by: (i) improving the quality of investments made by both public and private sectors; (ii) assisting countries and regions to establish roadmaps to achieve effective sector governance and reform using improved fisheries management tools; and (iii) aligning donor interventions

It is intended that PROFISH would improve sustainable livelihoods in the fisheries sector and to make concrete progress towards meeting the 2002 fisheries goals of the World Summit on Sustainable Development through three complementary activities: (i) ensuring sustainable fisheries initiatives are included in national plans and poverty reduction strategies; (ii) building national and regional consensus on pro-poor sustainable fisheries initiatives and priority activities to implement the Code of Conduct for Responsible Fisheries; and (iii) aligning and enhancing international assistance on fisheries and sustainable use of aquatic ecosystems and catalyzing implementation of agreed initiatives.

INTERNATIONAL SEAFOOD SUSTAINABILITY FOUNDATION

International Seafood Sustainability Foundation (ISSF) was formed in 2009 as a global, non-profit partnership among the tuna industry, scientists and WWF, the global conservation organization. The group states its mission is to undertake science-based initiatives for the long-term conservation and sustainable use of tuna stocks, reducing bycatch and promoting ecosystem health. ISSF has a Board of Directors which is advised by a Science Advisory Committee. The Foundation publishes a yearly report called The Status of the World Fisheries for Tuna which compiles the most recent science to determine the health of the 19 tuna stocks which support commercial fishing. In 2010, major retailers began unveiling sustainability policies that support the work of ISSF, including Walmart and Loblaw.

SEA SHEPHERD CONSERVATION SOCIETY

The Sea Shepherd Conservation Society (SSCS) is a non-profit, marine conservation organization based in Friday Harbor, Washington in the United States. The group uses direct action tactics to protect sealife. Sea Shepherd currently operates the vessels MY Steve Irwin, the Bob Barker, and the MV Gojira, and most of the group's recent activities take place in international waters. The group has a strong focus on public relations to spread their message via the media.

The organization was founded in 1977 under the name Earth Force Society by Paul Watson, an early member of Greenpeace, after a dispute with that organization over what Watson saw as its lack of more aggressive intervention. It has received support for its tactics against fishing, whaling, and seal hunting from quarters such as media personalities, while critics have called the direct action violent.

Operations have included scuttling and disabling whaling vessels at harbor, intervening in Canadian seal hunts, ramming other vessels, trying to temporarily blind or disorient whalers with a laser device, throwing bottles of foul-smelling butyric acid onto vessels at sea, boarding of whaling vessels while at sea, and seizure and destruction of drift nets at sea. Sea Shepherd claims that their aggressive actions are necessary as the international community has shown itself unwilling or unable to stop species-endangering whaling and fishing practices. Some governments and organizations have referred to them as terrorists. In January 2010, Jun Yamashita of the Japanese Fisheries Agencies stated on the effect of Sea Shepherd's actions on the Japanese whale hunting fleet "(...) the NGO's actions have kept the fleet from reaching its quota the last few years."

In 2008, Animal Planet began filming the weekly series Whale Wars based on the group's encounters with the Japanese whaling fleet in the Southern Ocean, a development which brought the group much publicity.

The predecessor organization of Sea Shepherd, the "Earth Force Society", was formed in 1977, after its founder, Paul Watson was ousted from the board of Greenpeace for disagreements over his direct action activism which clashed with their pacifist ethos. Watson soon left Greenpeace. Initially without funding and with only a small group of supporters, in 1978 Watson managed to convince Cleveland Amory, head of the British Fund for Animals to fund Watson's first vessel, the Sea Shepherd.

The first direct action undertaken by Sea Shepherd was against Canadian seal hunting in the Gulf of St. Lawrence in March 1979. Also in 1979, the group made headlines when they rammed a whaling vessel for the first time. Such acts continued with Sea Shepherd claiming responsibility for damaging or sinking multiple whaling vessels through sabotage or ramming. The group has attempted to intervene against Russian, Spanish, Norwegian, Icelandic, Makah, Faroese, and Japanese whalers in multiple campaigns around the globe. Setting a pattern that the group would keep up in later years, the group managed to scuttle a Portuguese whaling vessel, though the first Sea Shepherd was impounded and lost. Watson says that he used the money gained from selling the story rights to fund his next vessel.

After having spent the 1980s undertaking a variety of controversial and dangerous operations in support of various marine conservation aims, in the 1990s the group has been described as having undertaken a shift in their public attitude. Having previously argued primarily from an ethical viewpoint, from the 1990s, Watson's group now also started ascribing themselves law enforcement powers, using its interpretation of maritime and conservation law, to describe themselves as an anti-poaching agency. In some cases in the 2000s, they cooperated with official government efforts against maritime poaching, such as in Costa Rican waters, though the agreements often did not last long before conflict ensued.

Starting in September 2010, Sea Shepherd Conservation Society has positioned a crew in Taiji, Japan to monitor and report their annual dolphin slaughter.

Organization

Sea Shepherd is a non-government environmental organization and in the United States has a 501(c)(3) tax exempt status. About 80.8% of the organization's revenue is spent on its programs, while 8.9% of revenue is spent on administrative costs. Sea Shepherd is supported by private and corporate donations, lectures by Paul Watson, internet advertising, and grants. The group is operated by volunteers and a small paid staff. Watson says he is committed to keeping his organization small, and does not believe in spending money on fund-raising or recruitment. (though the group does fund-raise online and has received funds through a number of cooperation/donation agreements from other companies or organizations).

Sea Shepherd is governed by a board of directors. Currently, there are seven directors, including Watson. The organization has several boards of advisers, each addressing a particular area of expertise. These are the Scientific, Technical, and Conservation Advisory Board, which has 13 members including Earth First! founder Dave Foreman and Horst Klienschmidt, a former (2006) Deputy Chair of the International Whaling Commission; the Financial and Management Advisory Board, with three members; the Legal and Law Enforcement Advisory Board, with four members including Ian Campbell, a former Australian Minister of the Environment and Heritage (2004-07) whom whaling groups had previously accused of having inappropriate and close ties with the organization; the Animal Welfare, Humane and Animal Rights Advisory Board, with seven members including animal rights philosopher Tom Regan; the Media and Arts Advisory Board, with 15 members including Sean Penn and Martin Sheen; and a Photography Advisory Board with two members.

Activism

Sea Shepherd engages in conventional protests and direct actions to protect marine wildlife. Sea Shepherd operations have included interdiction against commercial fishing, shark poaching and finning, seal hunting, and whaling. The group has been active in intervening against fishing and poaching in the South Pacific, the Mediterranean, and in waters around the Galapagos Islands.

According to its mission statement, Sea Shepherd Conservation Society "uses innovative direct-action tactics to investigate, document, and take action when necessary to expose and confront illegal activities on the high seas". Those actions have included scuttling and disabling commercial whaling vessels at harbor, ramming other vessels, throwing glass bottles of butyric acid on the decks of vessels at sea, boarding of whaling vessels while at sea, and seizure and destruction of drift nets at sea. As of 2009, Paul Watson has said that the organization has sunk ten whaling ships while also destroying millions of dollars worth of equipment. Their practice of attacking and sinking other ships has led to reports of injuries to other sailors as well as the Sea Shepherd crew, including concussions and complications from chemical attacks.

Watson considers the actions of Sea Shepherd to be against criminal operations and has called the group an anti-poaching organization. Critics claim that Sea Shepherd's actions constitute violations of international law, while Watson has stated that Sea Shepherd believes that their actions constitute an attempt to enforce international conservation laws and international maritime law under the World Charter for Nature adopted by the United Nations . Australia has declared Japan's hunt in the Southern Ocean Whale Sanctuary to be illegal, federal court judge Jim Allsop has stated "there is no practical mechanism by which orders of this court can be enforced." The lack of official enforcement mechanisms in that law prompted the Society to adopt, without official sanction, what it sees as a law enforcement mission. A 2008 academic paper by researchers at Monash University concluded that the group "may be best categorized as a vigilante group, because they say they are seeking to enforce a legal status quo because of states' and the international community's inabilities or unwillingness to do so."

Watson left Greenpeace in 1977 after being expelled from the board of directors due to his confrontational methods. Since then, Greenpeace has criticized Sea Shepherd for the group's tactics, particularly regarding its interaction with whaling ships while at sea. The rival environmental group maintains Sea Shepherd is a violent organization whose tactics may endanger the lives of fishermen and whalers. Greenpeace has called Watson a violent extremist and will no longer comment on his activities. Greenpeace is also critical of the group on its website and state: "By making it easy to paint anti-whaling forces as dangerous, piratical terrorists, Sea Shepherd could undermine the forces within Japan which could actually bring whaling to an end". Both groups protest the Japanese whale hunts in the Southern Ocean but Greenpeace has a policy to not assist Sea Shepherd in finding the whalers. In his 2009 book, Whaling in Japan, Jun Morikawa states that Sea Shepherd's confrontational tactics have actually strengthened Japan's resolve to continue with its whaling program. According to Morikawa, Sea Shepherd's activities against Japan's whaling ships have allowed the Japanese government to rally domestic support for the program from Japanese who were otherwise ambivalent about the practice of hunting and eating whales.

Sea Shepherd has been criticized and sometimes physically attacked by people in several of the countries they protest against. In November 1988, Makah seized an inflatable boat belonging to the group and threw rocks at the Sea Shepherd's Sirenian in response to protests over their whale hunt. In March 1995, a mob of Canadian seal hunters stormed the hotel where members were staying. They fled while the mob ransacked their room. In 2005, Sea Shepherd crew were confronted by a threatening group of sealers while on the ice. In 2008, fishermen in the French islands of Saint Pierre and Miquelon cut the mooring lines of the Farley Mowat after hearing Watson make disparaging comments about the deaths of four seal hunters. In February 2010, pro-whaling demonstrators gathered outside the Australian

Embassy in Tokyo to protest the group. A political activist said that Sea Shepherd's actions were "absolutely racial discrimination against Japanese people." In response to the events of the sinking of the Ady Gil in January 2010, Glenn Inwood, whose firm handles public relations on behalf of the Japanese Institute of Cetacean Research accused Sea Shepherd of being "hostile eco-terrorists".

Governmental Response

In testimony on "The Threat of Eco-Terrorism" given to the a US congressional subcommittee in 2002, Sea Shepherd was the first group mentioned by an FBI official for having attacked commercial fishing operations by cutting drift nets. An earlier Canadian intelligence report on single issue terrorism" stated that "Watson and his supporters have been involved in a number of militant actions against whale hunting, driftnet fishing, seal hunting and other related issues" and mentions "activities against logging operations in Canada". The group has been accused of eco-terrorism by the Japanese government. Due to the 2008 operations against Canadian seal hunters, Danny Williams, Premier of Newfoundland and Labrador, called Watson a terrorist and said that Sea Shepherd was not welcome in the province.

Sea Shepherd has based many of its operations out of Australia with foreign crew members being able to travel in and out of the country on tourist visas. Tasmanian Greens and the Greens Senator Bob Brown, have endorsed and supported the Society in various ways, including advocacy within the Australian government and public endorsement of the group. When the Steve Irwin returned to Hobart in February 2009, Australian Federal Police seized film footage and the ship's logs, reportedly prompted by complaints from Japan. Brown demanded that the Australian Prime Minister, Kevin Rudd, order their immediate return, but a spokesman for the Home Affairs Minister said it was a federal police matter.

In October 2009, The Australian Immigration Department ruled that Watson and his First Officer, Peter Hammarstedt, must satisfy new good-character requirements to obtain business visas, requiring them to provide police references from the governments of the United States, Canada and Norway. Watson criticized what he considered a submission to Japanese pressure by the Rudd government. The Australian government responded by rejecting the idea that it was in some way delaying Watson, and on October 20, 2009 issued visas to Watson and Hammarstedt.

Paul Watson said to Discovery Channel the Dalai Lama sent a letter of support for Sea Shepherd's volunteers accompanied by a wrathful, scowling statue of the deity Hayagriva, which expresses compassion and determination in overcoming obstacles. In 2010 during a visit to Japan, the Dalai Lama said that while he agrees with the goal of stopping Japan from hunting whales, they should stop using violent methods to achieve that goal.

The ships of the fleet have flown the flags of different nations and the opinion of several governments that the vessels are engaged in inappropriate activities has several times led to registration issues for Sea Shepherd vessels. Canada, Belize, UK and Togo have revoked the registrations of various vessels. Both the MY Steve Irwin and MY Bob Barker ships now sail under Dutch flag leading to direct complaints by the Japanese government towards Dutch ambassadors. The Netherlands consequently considered revoking the registrations for both vessels but finally decided not to do so.

5

INLAND FISHERIES

Fish and other living aquatic resources of inland water ecosystems provide important services that are seriously undervalued. Inland fisheries and aquaculture contribute about 25% to the world's production of fish. In addition, many important estuarine and coastal fisheries are strongly linked to the ecological processes that occur in freshwater systems . The value of freshwater production to human nutrition and incomes is much greater than gross national production figures suggest. The bulk of production is generated by small-scale activities, with exceedingly high levels of participation not only in catching and farming, but also in processing and marketing. Inland fisheries are often critical to local food security. Most inland fisheries in the developing world are heavily exploited. While the fisheries are not necessarily overexploited in terms of gross production, individual species are often seriously overexploited. However, inland fisheries suffer greatly from environmental pressures, in particular deteriorating water quality and habitat. Many coastal and inshore marine systems are also affected by lower water quality and reduced availability of freshwater.

Competition for water and aquatic habitat is the most critical challenge facing inland fisheries in many countries . The need for water to support fish and fisheries can conflict with the needs of other sectors, in particular agriculture, in both water quality and flow requirements for sustaining aquatic habitat. Decisions on water management frequently do not take into account the impact on fish and fisheries and on the rural livelihoods of the populations that depend on them.

In part this is because inland fisheries are greatly undervalued in water management at local, national, and basin levels. Equally, there is a lack of knowledge of how to optimize ecosystem services, for example, through environmental flow and water productivity approaches that are needed to guide the allocation of sufficient water to sustain fish and fisheries.

Improving the consideration of fisheries in water management decisions requires better valuation methods and improved governance. Valuations need to pay more attention to nonformal values, especially those concerning livelihoods, food security, and biodiversity. Governance systems need to incorporate such values into cross-sectoral water management that recognizes the importance of ecosystem services. Decentralization may be a possible avenue toward these governance improvements but should be planned and implemented with care if equity in access to the resource, and its full development value, is to be fostered.

There are two broad challenges for fisheries production. The first is to sustain current levels of fisheries production and other ecosystem services through the provision of targetdirected environmental flows that sustain or restore the aquatic environment, including its diversity , and improved management of capture fisheries. The second is to increase current levels of fisheries production through the wider adoption of methods for enhancing and intensifying production, such as stocking and aquaculture, that require adequate quantities of clean water, suitable habitat, and appropriate management arrangements. These challenges will be more successfully addressed by building partnerships between fisheries and other interest groups concerned with water management, especially those engaged in water management for agriculture, which are also searching for more efficient ways to increase the overall benefits of water productivity to food security and poverty reduction.

WHAT INLAND FISHERIES CONTRIBUTE TO ECONOMIC AND SOCIAL DEVELOPMENT ?

Freshwater supplies are not only vital to river fisheries; they also sustain the fisheries of associated wetlands and influence estuarine and inshore marine fisheries.1 Although fisheries are usually nonconsumptive users of water, they require particular quantities and seasonal timing of flows in rivers and their dependent wetlands, lakes, and estuaries. There is, therefore, a tradeoff between the use of water for agriculture and the provision of water in the quality and quantity required for fish production and the other goods and services generated by inland aquatic ecosystems.

Fisheries and aquaculture from lakes, reservoirs, rivers, ponds, and wetlands contributed about 25% (34 million metric tons) of reported world fisheries production in 2003 (FAO 2004). However, catches in rivers and associated wetlands are easy to underestimate because the contributions of numerous fisheries on smaller tributaries and water bodies are generally overlooked (Coates 2002). Reported harvests from river fisheries alone have been shown to account for only some 30%-50% of actual catch (Kolding and van Zwieten forthcoming), and the contribution from inland fisheries is therefore believed to be significantly higher. In addition, the benefits of inland fisheries tend to accrue to local communities, in particular the rural poor, and their socioeconomic value is disproportionately high compared with other fisheries sectors such as high seas fisheries.

Despite the high productivity, water resources development planners give little recognition to freshwater-dependent fishery production or its ecological basis. Several factors contribute to this. One is the dearth of reliable data and scientific literature compared with that on industrial marine fisheries. The majority of freshwater fisheries are small-scale, spatially diffuse activities, and a significant part of the production is not commercialized or is marketed only through informal channels and is therefore not properly reflected in national economic statistics. As a consequence, these fisheries are often perceived as a low-value activity (Allan and others 2005). Aquaculture is better defined, although the reporting of fisheries that interface between enhanced natural fisheries and extensive aquaculture is less clear and makes the relative contribution of each sector difficult to assess. In addition, hydrological approaches to water management have tended to focus on in-stream quantitative flows, often ignoring the more important impacts on the quality and extent of adjacent wetlands.

The poor appreciation of the importance of the fishery sector has several consequences. It has exacerbated the lack of data, which has in turn hampered research and management (Misund, Kolding, and Fréon 2002) and may have biased policies and the allocation of national development resources away from fisheries. The contribution of fish to national GDP is underestimated in many countries, despite the importance to income and livelihoods. In turn, the legislative and policy frameworks for the management of inland fisheries and aquaculture development have either been absent or tended to focus on overexploitation as the primary issue when the priority need has been to improve environmental management. This impedes the role that these sectors can play at the scales of both national and local economies and in food security.

Contribution to the global economy

Fish is a highly traded commodity: roughly 33% of global fish output by value was traded across international borders in 2001 (Dey and others 2005), and it is now the fastest growing agricultural trade commodity on international markets. Since production remains more limited than for staple agricultural crops, and consumer demand for a healthier diet-often defined to include more fish-is increasing, the relative importance of fish is likely to continue to rise. In value terms the growth of trade in world fish products is greater than the increase in the net exports of other staple agricultural commodities such as coffee, bananas, rice, and tea (FAO 2002). In 2002 the value of world exports of fish and fish products increased to $58.2 billion-a 5% increase over 2000 (FAO 2004). At an estimated 8 million metric tons, the contribution of inland capture fisheries to total world fish production is small in comparison with marine capture fisheries and marine and inland aquaculture. Nevertheless, inland fisheries have sustained annual growth of about 2% worldwide (FAO 2002), and the potential for further increases in production is high in some systems (Kolding and van Zwieten forthcoming).

Contribution to the national economy

Inland fisheries, and related export and regional trade, can play a significant role in the economy of regions and countries. The sector contributes 7% to GDP in Cambodia (photo 12.1) and 4% in Bangladesh. In Africa inland fisheries provide employment and income for several million people. A recent estimate of employment and income for seven major river basins finds that in West and Central Africa alone fisheries provide a livelihood to more than 227,000 full-time fishers and yield an annual catch of about 570,000 tons with a first-sale value of $295 million (table 12.1; Neiland and Béné forthcoming).

The study also estimates that the total potential annual fisheries production for the region (about 1.34 million tons with an annual value of $750 million) is more than twice the estimated actual production.2 Freshwater and brackish water aquaculture also play a major macroeconomic role in some Asian countries, notably as a source of foreign exchange and employment.

Contribution to the Local Economy

Where small-scale inland fisheries or aquaculture has been supported and well managed, fish-related activities play a critical role in generating wealth and sustaining economic growth (Béné 2006) . For example, research in the Zambezi floodplain reveals that inland fisheries generate more cash for households than cattle rearing in most cases and more than crop production in some cases (table 12.2). In Sri Lanka recent economic valuations have put the value of fisheries at about 18% of total economic returns to water in irrigated paddy production (Renwick 2001). This capacity of small-scale fisheries to generate cash, however, is still poorly recognized by both academics and decisionmakers. In addition, because fishers and, to a lesser extent, fish-farmers, can access cash year-round by selling fish, fisheries provide a "bank in the water" for remote rural populations that lack access to formal financial systems. This contrasts with agriculture, where farmers have to invest and then wait for harvest before earning cash returns.

In some river basins recreational fisheries also contribute significantly to the local economy. In Europe, for instance, the inland recreational fishing industry has been valued as high as $25 billion a year (Cowx 2002). Increasing numbers of developing countries, such as Argentina, Brazil, Chile, India, and several states of the Zambezi River Basin, are also using part of their fishery resource for recreational fisheries to boost their local tourist economy.

EXOTIC FISHES

The initial development in fisheries was planting high altitude streams with trout for recreational fishing by the British. The attempts by Dr.Day in 1863 and 1886 and Mr.M.Ivor in 1867 to import and stock trout in Nilgiris had failed. In 1906 Mr.Wilson obtained 100 fingerlings from Ceylon of which only 27 survived and were released in Parsons' Valley stream. In 1907 be obtained ova from New Zealand and successfully hatched them. This rainbow trout (Salmo gairdneri) established well in the Nilgiris district. Tench of Doctor fish (Tinca tinca) and English carp (Carassius carassius) were introduced from Europe into Ootacamund in 1874. Mr.Wilson introduced Gourami (Osphronemus goramy) from Mauritius and Java into India 1916. 44 fingerlings of Mirror Carp (European Carp Cyprinus carpio) imported from Ceylon were stocked in the Crescent pond in the Ooty Botanical Gardens in 1939. By 1943, they had bred and figerlings were available for further stocking. Tilapia mosambica was introduced into Madras Presidency in 1952. The other fishes introduced were common carp (Bangkok strain) in 1957, silver carp from Hongkong to Cuttack in 1959 and grass carp from Japan to Cuttack in 1959 and 1969.

Fish Culture

Mr.Wilson was called the founding father of inland fisheries development in the south. In memory of him, the fish farm at Ooty was named as "Wilson fish farm". The credit of establishing trout culture in the Nilgiris on a firm and scientific basis goes to Mr.Wilson. a trout hatchery at Avalanche in Nilgiris district was constructed in 1907. In 1909 a hilsa hatchery was established a Lower Anicut on the Coleroon River. Mr.Wilson succeeded in artificially fertilizing Hilsa ilisha eggs but the hatching was not successful. Dr.Sundara Raj took charge of the experiments in 1915.

Larvicidal Fishes

Mr.Wilson took special initiative in breeding and supplying of larvicidal fishes for mosquito control. Ippur fish farm in Nellore district was established in 1917 mainly to breed larvicides in large numbers and distribute throughout the Presidency. In order to have extensive experiments, a fish farm at Praema near Nandyal was established in 1927.

Reservoir Fisheries

Tamilnadu is pioneer in reservoir fisheries development. Even before the completion of the construction of the Stanley Reservoir (Mettur Dam) in 1934 pre-impoundment stocking of Catla was carried out. The cardinal principles of pre-impoundment survey, stocking, conservation and management were adopted. Every reservoir had a fish seed farm attached to it. Research work on the water quality and the productivity of the reservoirs was carried out. The evolution of "Rangoon" nets (vertical walled surface gill nets of different mesh size) had also its beginning in Mettur and found its way into other reservoirs.

Madras Rural Pisciculture Scheme

Mr. K.H.Alikunhi initiated research work on inland fisheries under the ICAR sponsored "Madras Rural Pisciculture Scheme" in 1942 and hundreds of rural fishery demonstration ponds sprung up all over Madras Presidency. This was the first extension project translating the research findings into production programme and was the forerunner of the very successful Fish Farmers Development Agency (FFDA) scheme initiated in 1973-74 by GOI. In the mid forties, Madras Fisheries had the distinction of starting a hydrobiological station in Chetpet, Madras.

India ranks second in inland fish production

India produces about 6.57 million metric tonne fish every year. The inland-sector, which has a growth rate of 6 per cent, contributes around 55 per cent of it.

According to S Ayyapan of Indian Council of Agricultural Research, India is home to more than 10 per cent of global fish biodiversity with 2200 species of fish and shellfish in the marine and inland waters.

"With water resources in terms of 29,000 km of rivers, 3.15 million hectares of reservoirs, 2.35 million hectares of ponds and tanks and 0.2 million hectares of floodplain wetlands, the potential production levels are estimated at over 4.5 mmt annually," Mr Ayyapan said.

"Rivers are sources of precious germplasm and breeding ground for a host of fish species. Reservoirs and floodplain lakes offer an opportunity for enhancing fish production," Mr Ayyapan added. He stressed the need for community intervention in stock enhancement and resource management. "For example, ponds, tanks and water harvesting structures are being put to aqua-culture practices, integrating a number of other farming systems".

Carps contribute as much as 87 per cent of the total aquaculture production,"points out Mr. Ayyapan. "This is achieved by combining mainstay culture practices with strong traditional knowledge and by applying scientific strategies in management"

With immense possibilities for both open water fisheries and fresh water aquaculture, India ranks second only to Japan in inland sector fish production. India produces about 6.57 million metric tonne fish every year. The inland-sector, which has a growth rate of 6 %, contributes around 55 % of it.

According to S Ayyapan of Indian Council of Agricultural Research, India is home to more than 10 % of global fish biodiversity with 2200 species of fish and shellfish in the marine and inland waters.

GEOGRAPHICAL LIMITATION OF INLAND FISHERIES

This chapter uses a GIS-assisted approach to illustrate some of the geographic constraints on enhancements. The perspectives described are global and continental rather than regional and national.

The underlying objectives are to:

- Illustrate that enhancement potential is finite rather than limitless;
- Show that some constraints on enhancements can be treated in a geographic context; and
- Demonstrate that GIS is a useful tool to assess enhancement potential and to contribute to strategic planning for enhancements.

Materials and Methods

In overview, geographically referenced digital data sets that were originally created mainly for global climate change analyses were assembled on a variety of topics related to geographical aspects of inland fishery enhancements. World Wide Web sites were a source of much of the data and documentation. Analyses were made at the FAO GIS Centre using ARC/GRID and IDRISI GIS softwares. Complementary data were drawn from the literature.

Geo-referenced digital data sets

Unless otherwise mentioned, all original data sets were grids of cell size of 1 1 degree latitude and longitude. All maps were converted to the Flat Polar Quartic projection in order to extract surface area estimates for the total global land surface (less Antarctica).

Below is a brief account of how the data sets were developed and described by their respective original authors.

These data were obtained by superimposing a template of varying size on 1:1 million-scale maps such that 100 points were sampled at regular spacing in each 1 1 degree grid cell. Therefore, the values obtained for perennial lakes and for swamps, marshes and other wetlands can be interpreted as an estimate of the percentage extent of that type of water body within the 1 degree cell.

The stream frequency counts made for the Perennial Rivers map have to be interpreted with caution because they represent the number of template intersections out of a possible 100 underlain by "blue-line features" that correspond to rivers and streams.

The IIASA Climate Database (Leemans and Cramer, 1990) consisting of 12 monthly values of precipitation, derived from data collected from 1930 to 1960, then selected and smoothed to fit 0.5 degree grids, was used to map mean annual precipitation. Mean monthly air temperature values were taken from the same source.

Average Annual Methane Emission from Animals (Matthews and Fung, 1988) was used as a surrogate for animal manure as a fertiliser in small water bodies. The statistics on animal population densities among 9 domestic types were obtained from FAO and other sources. Animal numbers were distributed across the one-degree latitude/longitude grids including sub-national boundaries for Australia, Brazil, Canada, China, India, the USA and the former USSR. Published estimates of methane emission by each type of animal were used along with the population numbers to estimate methane emission as kg/km2/y.

Cultivation Intensity from the Matthews Dataset (Matthews, 1983) was used as a surrogate for the availability of crop by-products as natural fertilisers. This data set draws on approximately 100 published sources complemented by a large collection of satellite imagery. Cultivation intensity defines the areal extent of presently cultivated land in the 1 1 degree cells in 5 classes ranging from 0, 1-20% up to 76-100%. The National Institute for Geographic Information Analysis (NCGIA) data set (Tobler et al., 1995), originally on 5 are minute grids, was used to establish human population numbers.

THE NILE TILAPIA AS THE BENCHMARK SPECIES

The Nile tilapia (Oreochromis niloticus) has been chosen to generalise the temperature-related limitations on enhancements of tropical freshwater food fishes. It has been used in this role previously by Kapetsky (1994) and by Kapetsky and Nath (in press). It serves well as a "benchmark" species because it is well known in fisheries in Africa, because of its wide geographic distribution elsewhere and due to its extensive use in aquaculture world wide.

GIS analyses

General aspects

Flow charts outline the main analytical steps used to identify standing water bodies in which there is continuous growth of Nile tilapia that are also proximate to markets for fish (Fig. 1) and to locate areas with small water bodies in which there is continuous growth of Nile tilapia and natural fertilisers as an enhancement (Fig. 2).

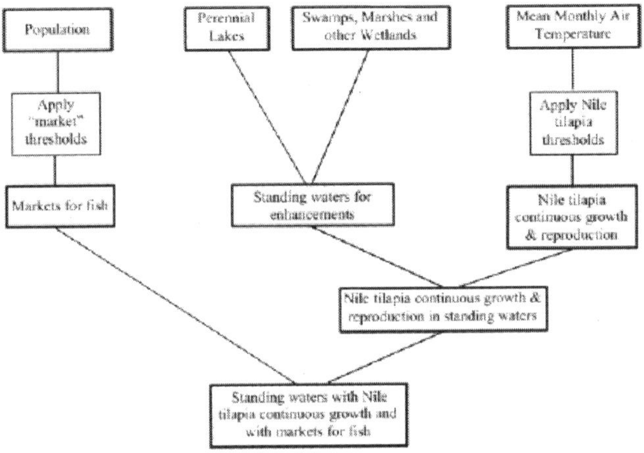

Figure 1. Flow chart of GIS analyses to produce standing waters that support Nile tilapia continuous growth and that have nearby markets for fish.

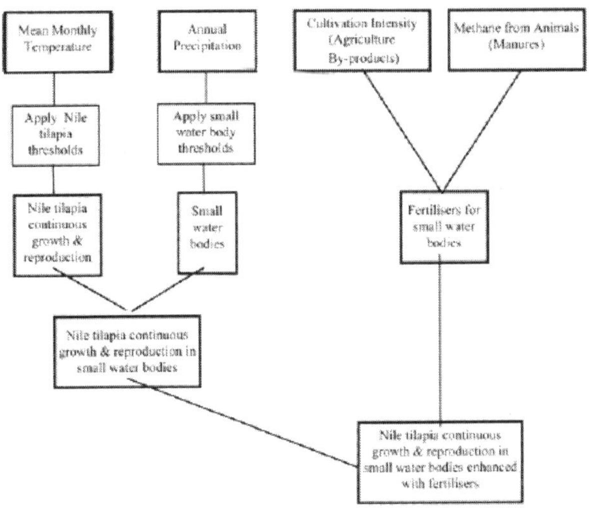

Figure 2. Flow chart of GIS analyses to produce Nile tilapia continuous growth and reproduction in small water bodies enhanced with natural fertilisers.

CONTINENTAL MASK BASED ON FAO INLAND FISHERIES STATISTICAL AREAS.

Relative continental surface area estimates were extracted using a mask based on the country groupings used by FAO to delineate Major Fishing Areas for Statistical Purposes (FAO, 1996) and a similar mask that included large inland water bodies was used for display of the results.

Standing waters

Standing Waters were derived simply by adding Perennial Lakes and Swamps, Marshes and Other Wetlands. Because of the way that the original data were obtained and mapped, as explained above, this was a straightforward process that resulted in 100 classes with each class expressing the percent of each grid cell containing standing waters.

Continuous growth and reproduction of Nile tilapia

The IIASA mean monthly air temperature data sets were manipulated and reclassified to map growth, reproduction and distribution of the Nile tilapia in the following way after Kapetsky (1994), using thresholds similar to those of Coche et al. (1996):

Table 1. Classification of growth and reproduction of the Nile tilapia in relation to temperature.

Class	Temperature Regime	Interpretation
1.	t=<14 C	Unable to grow and reproduce
2.	Intermediate	Growth and reproduction discontinuous
3.	t>=22 C for 12 months	Continuous growth; reproduction possible throughout the year

For simplicity and because of the low resolution of the study, it was assumed that mean monthly air temperature is an acceptable surrogate for water temperature.

Continuous growth of Nile tilapia in standing waters

Enhancements in standing waters can include stocking, net and cage culture and fisheries in blocked-off arms.

The focus herein is on the locations where there is good potential for enhancements using the Nile tilapia. For this reason only areas where growth is continuous and reproduction is possible throughout the year (Class 3, Table 1, above) were carried forward for further analyses.

Standing Waters were combined with the Nile Tilapia Continuous Growth Regime and interpreted for enhancements in the following way (Table 2):

Table 2. Classification of the continuous growth regime of the Nile Tilapia in standing waters.

Standing Waters per Grid Cell	Class Nile Tilapia Continuous Growth	Interpretation for Enhancements
<1%	0	Unsuitable
1-25%	1	Few opportunities
26-50%	2	Moderately few opportunities
51-75%	3	Moderately many opportunities
76-100%	4	Many opportunities

Continuous growth of Nile tilapia in standing waters and population

Enhancements in large standing waters will include the sale of most, if not all, of the catch locally. Therefore, market outlets for the catch are of importance. No direct measure of market potential was available at the global scale of this study. Therefore, population per grid cell was used as a surrogate. The following classification, thresholds and interpretations were made (Table 3):

Table 3. Population density interpreted as market opportunities for fish.

Class	Population (density/km2/grid cell)	Interpre tation of Market Opportunities
1.	<1	None
2.	1-10	Few
3.	11-100	Moderately few
4.	101-1000	Moderately many
5.	>1000	Many

The Continuous Growth of Nile Tilapia in Standing Waters and Population were combined. This classification assumes that markets are important even where there is a relatively low density of standing waters.

In other words, it is important to have markets even where the output from fisheries may be relatively low. It also assumes that where market possibilities are relatively few, but standing waters with continuous growth of Nile tilapia relatively high, fish might be transported some distance to markets.

Small water bodies

The IIASA 12 monthly mean precipitation data set was summed to provide mean annual precipitation. Then annual precipitation was interpreted in relation to the distribution of small water bodies using the same thresholds as developed for Africa by Kapetsky (1994) (Table 5). In short, data from a number of African countries indicated that with rainfall at 1200 mm/y or greater water is ample to keep ponds filled all year and that at 1000 mm/y water is usually sufficient; however, in the range 700 to 1000 mm/y special precautions may have to be taken including siting in relatively large watersheds and construction of relatively high embankments.

Below 700 mm/y ponds may dry out, or the high cost of construction, or low yields caused by seasonally highly varying volumes of water, may make them uneconomic, unless there is an ample ground water or perennial surface water supply. The risk relates to the possibility that fish production, or cost of production, will be affected by water availability and that fish production risk increases with decreasing water resources.

Table 4. Occurrence of small water bodies in relation to annual rainfall (after Kapetsky, 1994).

Class	Annual Rainfall (mm)	Interpretation
1	<700	Insufficient; High risk
2	700-999	Sufficient with cost-adding modifications; Moderately high risk
3	1000-1199	Sufficient with little or no cost-adding modifications; Low risk
4	>1199	Ample; Very low risk

Continuous growth of Nile tilapia in small water bodies

The occurrence of Small Water Bodies was combined with the Nile Tilapia Continuous Growth Regime to make "Nile Tilapia Continuous Growth in Small Water Bodies". Because there is only one class for Nile tilapia continuous growth, the interpretation is the same as that shown in Table 4 except that the occurrence of small water bodies is now limited to those in which continuous growth of Nile tilapia is possible.

Fertilisers as small water body enhancements

Cultivation Intensity as a surrogate for agriculture by-products was combined with Methane from Animals as a surrogate for animal manures to constitute the availability of natural fertilisers as enhancement inputs for small water bodies. Cultivation Intensity was weighted more heavily than Methane because agriculture by-products are more likely to be used as fertilisers than is manure (A. Tacon, pers. comm.). The resulting map is called "Natural Fertilisers".

Continuous growth of Nile tilapia in small water bodies with natural fertilisers

Nile Tilapia Continuous Growth in Small Water Bodies was combined with Natural Fertilisers to make a map of the availability of natural fertilisers to small water bodies in which there would be continuous growth of Nile tilapia (Table 7). The risk referred to below is the same as that set out in Table 5.

Table 5. Classification of the combination of Cultivation Intensity with Methane from Animals to derive Natural Fertilisers.

Cultivation Intensity (%)	Methane from Animals(kg/km2/y)			
	1-10	**11-20**	**21-30**	**31-40**
1-20	1	2	3	4
21-50	1	2	3	4
51-75	1	2	3	4
76-100	2	3	4	4

Table 6. Classification of the combination of the continuous growth of Nile Tilapia in small water bodies with natural fertilisers.

Continuous Growth of Nile Tilapia in Small Water Bodies (Classes)	Relative Quantities of Natural Fertilisers (Classes)			
	Low	**Moderately Low**	**Moderately High**	**High**
High Risk (1)	0	0	0	0
Moderate Risk (2)	1	1	2	3
Low Risk (3)	1	2	3	4
Very Low Risk (4)	3	3	4	4

In this combination greater weight was placed on lower risk in small water bodies than on availability of fertiliser as it was reasoned that a dependable supply of fish is more important for food security than one that may be higher in quantity due to fertiliser enhancements, but less dependable in supply.

RSULTS

Interpretation

Care in interpretation is required because results mainly are expressed in relative terms and given in relative quantities per grid cell, not in absolute area terms. The continent-wide estimates of area are likely to be underestimates because the continental boundaries of the "mask" used to extract the grid cell counts by continent did not spatially conform exactly to the near-boundary grid cells of the data files. Therefore, some grid cells were lost. This does not affect the global estimates of area because no mask was needed to derive them.

What quantity of water resources is available to be enhanced on a global and continental basis?

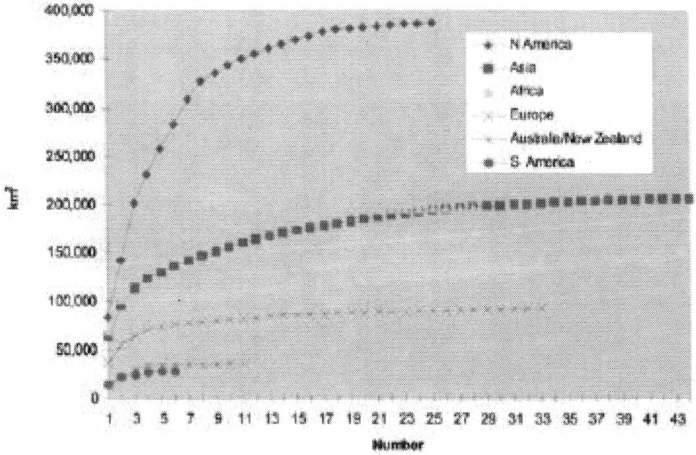

Figure 3. Areas and numbers of large lakes by continent.

This estimate is based on data originally extracted from 1:1,000,000-scale maps that, in theory, could include water bodies as small as 1 km2. In comparison, using data tabulated by Budyko et al. (1977), total large lake area is about 0.95 million km2 distributed among the continents as shown in figure 3. This tabulation includes lakes generally larger than 100 km2, thus accounting for the difference between the large lakes and perennial lakes estimates extracted from the digital map.

Swamps, marshes and other wetlands amount to some 4 million km2 about 3.8% of the global surface.

The grid cells containing more than one perennial river are widely distributed albeit at relatively low frequencies in most areas. Summing the river channel lengths tabulated by Budyko et al. (1977) provides another quantitative measure of rivers, some 269,000 km globally, that are distributed by continent as shown in Figure 6.

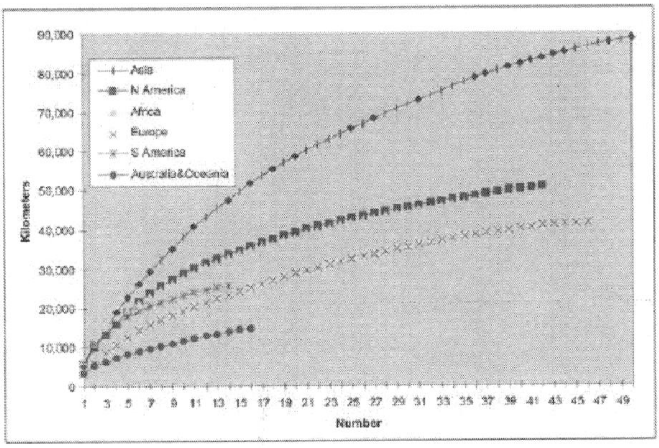

Figure 4. Main river channel lengths and numbers by continent.

What are the prospects for an increase in enhanceable surface area from new reservoirs?

New water surface for enhancements can come only from restoration of degraded aquatic systems and from the creation of new water bodies. Only the latter is dealt with herein because of the scarcity of synoptic data on the former. Total global reservoir area is elusive. The organisation that monitors reservoir construction and other characteristics, the International Commission on Large Dams, does not include surface areas in its tabulations and therefore the global information presented herein is based on numbers of dams that are greater than 15 m in height. These numbered 36,235 in 1986 (Anon., 1989) of which 64% were in Asia, 21% in the Americas and 12% in Europe.

Globally, most of the existing reservoirs were created after World War II. Most reservoirs are in Asia and China has been at the forefront of dam building often explicitly including engineering for fishery enhancements at the outset of the design stage (Li and Wu, 1995), e.g. blocking of coves for fish hatcheries, earthmoving to facilitate fishing. New dams greater than 15 m in height under construction in 1994 (Anon., 1997) numbered 948 and the continental distribution is shown in Figure 5. Asia is foremost, but China ranks fourth after Japan, Turkey and the Republic of Korea, suggesting that the rate of creation of new, large reservoirs is decreasing there.

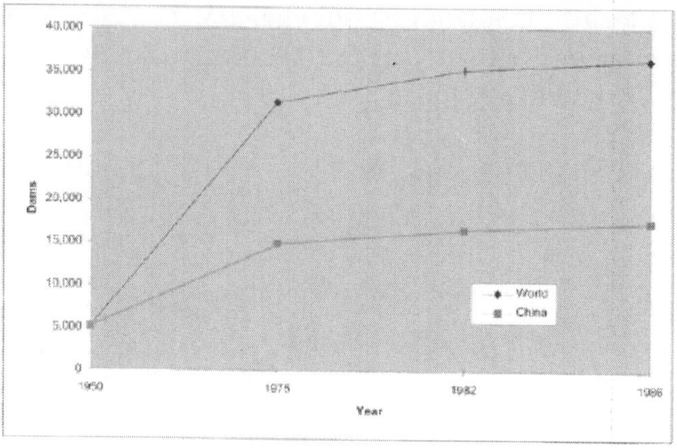

Figure 5. Dams higher than 15 m.

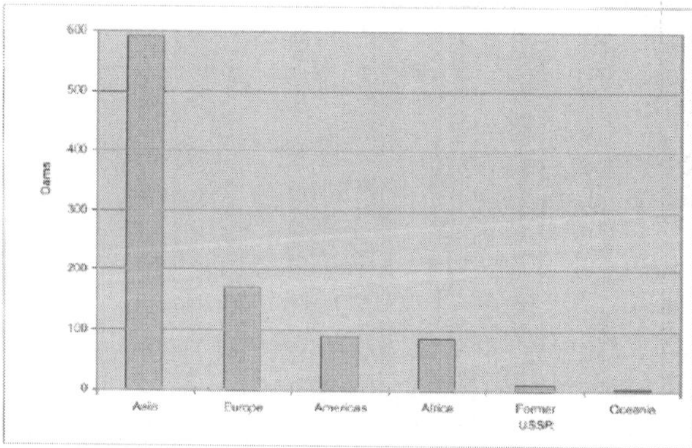

Figure 6. New dams under construction in 1994 higher than 15 m.

In Africa, there was relatively little reservoir surface area before 1955, with peak large reservoir construction from the late 1950s to the early 1970s (Fig. 6; data from FAO Land and Water Development Division). The trend appears to be for fewer and smaller reservoirs for the near future for Africa. There, as well as elsewhere in the world, the environmental consequences and social impacts of large reservoirs are increasingly being called into question (e.g. Covich, 1993).

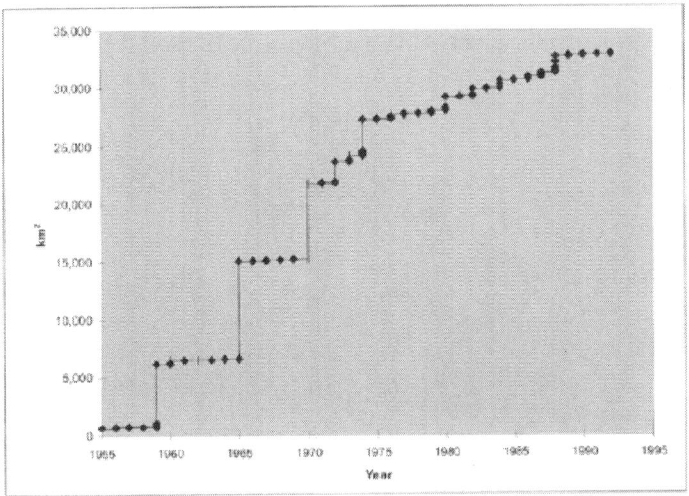

Figure 7. Cumulative reservoir area in Africa.

Expectations for significant increases in water surface for enhancements in new reservoirs in the near future are not justified except for some Asian countries, and a few countries each among Europe, Africa and the Americas according to the data examined. However, the expansion of irrigated cropland to meet food security needs in relation to the increasing problem of water scarcity (Kulshreshtha, 1993) may produce an increase in reservoirs as attempts are made to meet the demand for water. This will put a premium on fishery enhancement technologies that combine species that can achieve good growth, probably in short growing seasons, and enhancement approaches that are efficient in widely fluctuating environments.

HOW ARE ENHANCEMENT PROSPECTS AFFECTED BY CLIMATE?

As indicated above, enhancement prospects are largely determined by the present and future availability of water surface. Another basic consideration is the temperature regime as it relates to growth and reproduction of fishes that are employed for enhancements. This can be illustrated by using the ubiquitous Nile tilapia as an example of a tropical food fish. Specifically, the first question is "Where globally could Nile tilapia be used for enhancements given a temperature regime in which growth and reproduction are continuously possible throughout the year?".

Globally, the areas of the world where good performance could be expected include about 18% of the land surface. On a continental basis South America is most favoured with about 47% of its area in the regime in which growth and reproduction can be continuous, followed by Africa (Fig. 7).

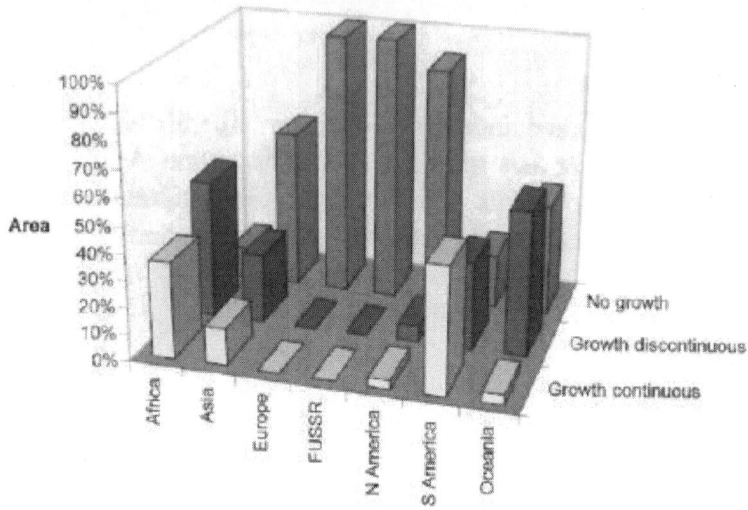

Figure 8. Growth of Nile tilapia in relation to temperature.

The next question is "Where does the continuous growth regime for Nile tilapia coincide with a relatively high incidence of standing waters?". Enhancements possible in standing waters encompass nearly all of the varieties including stocking, pen and cage culture. Standing Waters have been obtained by adding Perennial Lakes (Fig. 8) and Swamps, Marshes, and other Wetlands.

Although South America rated high for a favourable growth regime for Nile tilapia, when standing waters are taken into account, only about 11% of the continental area combines the continuous growth regime with a density of standing waters of 1-25% per grid cell. Occurrences of this regime at higher densities of standing waters are in less than 0.01% of the continental area.

When food fisheries are enhanced, usually most of the fish are for sale. Therefore, prospects for enhancements are affected by availability of markets. Using population as a surrogate for market size and combining it with standing waters in which there is continuous standing waters with many or moderately many fish market opportunities are, relatively, very scarce. From a continental perspective, Africa followed by Asia and South America have most of the potential.

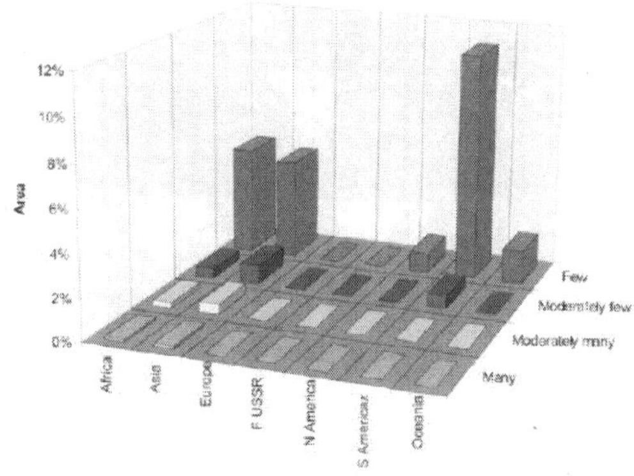

Figure 9. Opportunities for enhancements in standing waters in which the growth of Nile tilapia is continuous.

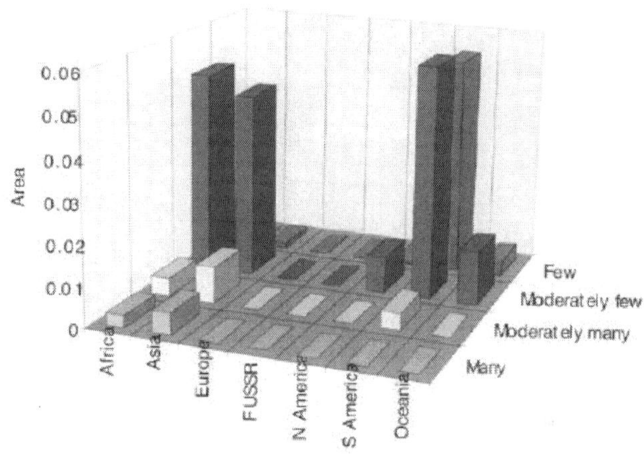

Figure 10. Standing waters with markets for fish and continuous growth of Nile tilapia.

ENHANCEMENTS IN SMALL WATER BODIES

The potential for enhancements in relatively large standing water bodies has been dealt with in previous sections; however, in certain combinations of water-holding soil types and seasonal rainfall regimes numerous small water bodies have been constructed to serve as community water supplies for domestic consumption, for watering livestock and for small-scale irrigation. These small multi-purpose water bodies are believed to have good fishery potential, if properly managed, and have the advantage that they often coincide with relatively poor farming communities in areas that are marginal for rain-fed agriculture so that fish could become an important contributor to food security. Thus, the question being posed in this section is "What is the occurrence of small water bodies in which there also are continuous growth prospects for Nile tilapia?".

A major problem for the implementation of enhancements is the inventory and characterisation of small water bodies because such knowledge is very often lacking (Haight, 1994; also see Verheust, this volume). Therefore, for the purposes of illustration herein it was reasoned that the occurrence of small water bodies can be estimated globally on the basis of mean annual rainfall (Table 5) in the same way as was carried out for ponds in Africa by Kapetsky (1994). With these thresholds, about 60% of the global land mass is too dry for perennial small water bodies, or volumes would fluctuate widely. About 11% could support small water bodies with cost-adding modifications (moderately high risk), in about 6% water would be sufficient with little or no cost-adding modifications (low risk) and some 23% would have ample water (very low risk) (Fig. 9). On a continental basis, Europe possesses the largest relative amount of territory favouring small water bodies, about 40%, while Africa, South and North America are similar with from 18 to 20% and the former USSR and Oceania have less of this capacity.

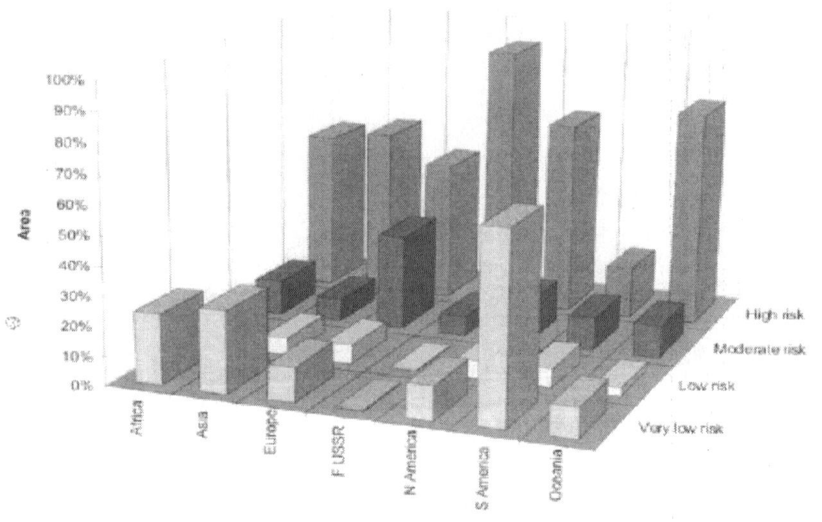

Figure 11. Occurrence of small water bodies in relation to annual rainfall.

Combining the Continuous Growth Regime of Nile Tilapia with the occurrence of small water bodies shows that only about 1% of the land surface encompasses conditions for continuous growth of Nile tilapia and low risk for small water bodies while an additional 12% combines the continuous growth regime with very low risk for small water bodies. Continentally, South America is most favoured with about 46%, Africa is next with considerably less, about 24%, and Asia possesses about 12%.

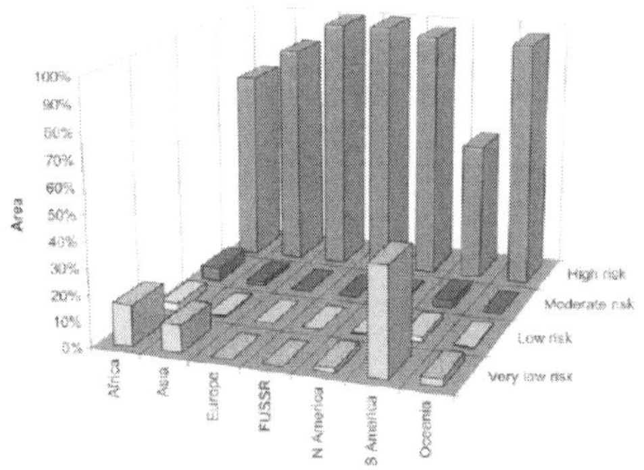

Figure 12. Small water bodies with continuous growth of Nile tilapia.

HOW ARE ENHANCEMENT PROSPECTS AFFECTED BY AVAILABILITY OF NATURAL FERTILISERS?

Fertilisation is an enhancement that has been put to good use most notably in China. Elsewhere, there will be more and more interest in fertilisation in situations where high stocking rates alone are unable to produce greater yields because the full natural productive capacity of the aquatic system is already being realised. An obvious question then is "Where are there prospects for fertilisation of small water bodies using crop by-products and manures as inputs and also where is continuous growth of Nile tilapia possible?". The available surrogate for agriculture by-products is cultivation intensity and that for production of animal manures is the production of methane from animals. These have been combined to show the distribution of natural fertiliser inputs for small water bodies.

Methane from animals (kg/km2/y) interpreted as availability of manures as fertilisers.

This analysis suggests that only about 19% of the global land surface could provide fertiliser inputs at from relatively low to relatively high quantities. Europe stands out as having a significantly greater quantity than other continents, with about 48% of its area in this category.

In contrast, only 5 to 24% of the land area of the other continents could provide such fertilisers. Although the apparent "natural fertiliser capable" areas of most continents are small, the areas that could produce intermediate to relatively high quantities are in Europe, Asia and North America. Africa, in contrast, has most of its fertiliser-capable land in low quantities of natural fertiliser output.

Globally, only about 4% of the surface encompasses continuous growth conditions for Nile tilapia, sufficient to ample water to maintain small water bodies (i.e. moderately high to high certainty of small water bodies) and moderately high to high quantities of natural fertilisers. Africa leads in this regard with about 11% of its area combining these three conditions followed by Asia with about 4%, whereas the other continents possess only very modest areas, or lack this capability altogether.

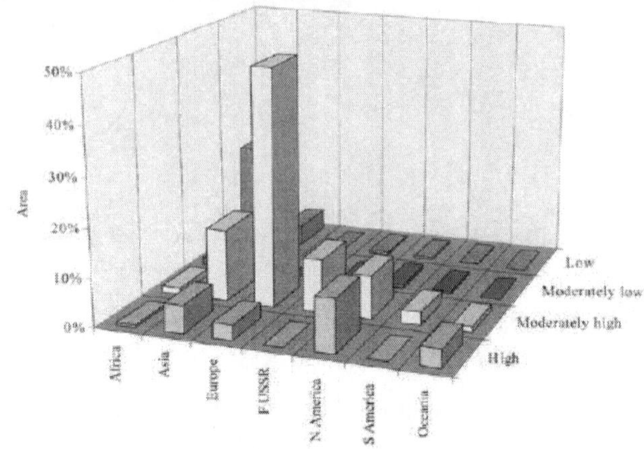

Figure 13. Availability of natural fertilisers for small water body enhancement.

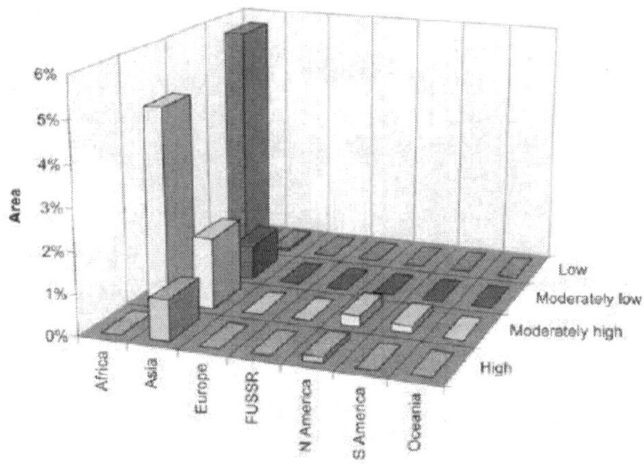

Figure 14. Availability of natural fertilisers and certainty for small water bodies with continuous growth of Nile tilapia.

DISCUSSION

Interpretation

This is a global study of the constraints on enhancements, carried out at low resolution and using only a few basic determinants of potential.

However, the approach does suggest how a similar approach could be used with higher resolution data at national levels. Some of the quantitative results, because of the resolution and sometimes because of the use of surrogate data, have to be considered as indicative and such cases have been so qualified.

Inland waters for enhancement

Areas for inland fishery enhancements include about 1.7 million km2 of lakes and about 4 million km2 of swamps, marshes and other wetlands. These are equivalent to about 1.4% and 3.8% of the global surface, respectively. Large reservoirs numbered some 36,235 in 1986 and those under construction in 1994 were 948. Thus, the growth of new reservoirs is not rapid at present and is very localised among relatively few countries. Uncounted small multiple purpose reservoirs that are found in some rainfall regimes offer additional possibilities, but technical, financial, social and administrative impediments have to be overcome before their enhancement potential can be realised. Perennial rivers, mainly at a relatively low frequency per unit of area, are widespread. Main river channels sum up to some 180,000 km.

Looking at enhancement potential purely on a water resources basis by continent, North America, Africa and Asia are favoured in lake area. Asia has the most abundant reservoir numbers, followed by the Americas and Europe. Small water body potential appears to be high in Europe, Africa and the Americas. Swamps, marshes and other water bodies are well distributed among the continents. Asia, North America and Africa dominate in terms of rivers when measured in terms of channel length.

Constraints on enhancements

Water body size in some respects is inversely related to enhancement potential. For example, much of the surfaces of large inland standing waters may be out of reach of enhancements other than for introductions and stocking. Open waters of large lakes and closed and remote areas of large swamps can not be effectively fished by artisanal or recreational fishermen. Other kinds of enhancements, such as pens, are shore-bound. For the present, cage farms require shelter and easy access. Thus, surface alone has to be discounted as an indication of enhancement potential. Shoreline length could be a better overall measure and shoreline development could better indicate the relative amounts of shelter for cages and pens.

Enhancement opportunities depend not only on the total area available, but also on the type of water body and its inter-annual and seasonal variability in surface area. For example, as seen above, Africa figures as a continent with important water resources with potential for enhancements.

However, Kapetsky (1995) noted that the hallmark of African fisheries potential is annual variability. Systems that fluctuate seasonally and inter-annually in surface area--reservoirs, shallow lakes, swamps and river floodplains--account for about 57% of the total water surface area there. Turning to small water bodies, it has been shown that they are most likely to occur in areas with relatively low and annually varying rainfall that may cause them to dry out completely, or to provide unfavourable conditions for fish production. They also present administrative and financial challenges for the implementation of enhancements due to their abundance, small size and often scattered distribution.

There are many constraints on the implementation of enhancements in the world's water bodies, but climate is the most basic. For example, most of the large lake area and that of swamps, marshes and other wetlands are found in temperate or cold temperate regions of the northern hemisphere where aquatic production is low.

Further, it has been shown that only about 18% of the world's area could support the continuous growth and reproduction of the Nile tilapia. In so far as the Nile tilapia is indicative of temperature-growth regimes of other tropical fishes, this area roughly defines the amount of surface that could provide the equivalent, in fish farming terms, of two harvests per year. The remaining are discontinuous growth areas that can produce the equivalent of about one harvest per year, or less. Moving along the continuum from capture to culture fisheries, enhancement using fertilisers will be increasingly important to obtain higher yields. In this regard, the prospects for fertilisation as an enhancement method in Africa do not appear too good for the short term. Apart from the apparent low level of natural fertiliser availability from the fertiliser-capable areas, Africa accounted for only 2.9% of the world manufactured fertiliser consumption in 1994/95 (FAO, 1997).

This contrasts with some 48% of the world's fertiliser consumption among the developing countries of Asia. Only 10 countries in Africa are consumers and two-thirds of the total manufactured fertiliser consumption occurs in only 3 countries. Production capacity exists in only 8. Therefore, in place of naturally produced fertilisers, most African countries would have to import manufactured fertilisers in order to increase fish production in this way, an alternative that seems unlikely for the present.

APPROACHES TO THE ESTIMATION OF ENHANCEMENT POTENTIAL

For purposes of illustration, global data sets created for global change analyses have been employed herein to provide a world-wide perspective on constraints on enhancements; however, studies to assess enhancement potential are best carried out at a national level in order to take advantage of varied and high-resolution digital data sets and to better account for social, cultural and economic variations from one part of a country to another. For example, in place of population density used as a surrogate for fish markets in this study, Kohler (1996) has used census data to combine known consumer trends with an area's demographics to make an initial market assessment for aquaculture products. This raises an important point, that is, for GIS-assisted studies, much the same data are used for assessments both of enhancement potential and of aquaculture potential and the additional time required to do both together rather than one alone is not great. Therefore, both should be carried out in concert.

Digital data with sufficiently fine resolution to be useful for national-level studies of enhancement potential are becoming increasingly available. Among these are the Global Land Cover1 at 1 km resolution, Topographic and Climate Data Base for Africa2 at 3 are minute resolution, temperature and precipitation for Latin America and the Caribbean at 10 are minute resolution3, World Water and Climate Atlas4 at 2.5 km resolution and Drainage Basins5 at about 5000 km2.

For example, Kapetsky and Nath (in press) have used gridded temperature data sets to predict the number of crops possible per year for four fish species on each 9   9 km grid cell of Latin America and they have used the precipitation data sets to predict water losses from ponds at the same resolution. Both of these approaches are directly transferable to small reservoirs and the former to all water bodies for which monthly temperature regimes can be estimated.

CONCLUSIONS

At first glance, the opportunities for inland fishery enhancements seem nearly limitless; however, climate and water availability act to make this area quite small relative to the continental land masses.

Other limitations such as degradation of the aquatic environment and social and economic aspects of access, finances and technical skills, particularly in some developing countries, restrict prospects for rapid implementation of enhancements. Expectations for significant increases in water surface for enhancements due to new reservoirs are not justified except for some Asian countries, and a few countries each among Europe, Africa and the Americas. Therefore, most new enhancements will have to be implemented in existing water bodies. The approach to estimate enhancement potential and the data required to do so are very similar to those employed to estimate aquaculture potential. Therefore, for purposes of economy, both should be carried out together. However, as enhancement potential depends fundamentally on water quality and water quantity, there is ample justification for studies of enhancement potential, or of loss of enhancement potential, from an international river or lake basin viewpoint.

6

METHOD OF INLAND FISHERIES

States and users of living aquatic resources should conserve aquatic ecosystems. The right to fish carries with it the obligation to do so in a responsible manner so as to ensure effective conservation and management of the living aquatic resources.

Responsibility for management Inland waters differ from most marine environments in that in that the primary responsibility for decisions affecting the environment and it living aquatic components rarely lies with the fishery authorities but with some other agency. Fisheries interests are therefore peripheral to the policy making and allocation process in most countries. The following remarks expand on the consequences of this line of responsibility for conservation and management of the fishery.

The value of aquatic ecosystems lies in the sustained net benefits derived from the many goods and services they supply including various ecological functions, products for direct and indirect human consumption, energy, aesthetic and recreational benefits, and assimilative capacity of residues of human activities. Trade-offs between the net benefits of one use vis-à-vis those that can be derived from other uses are usually necessary. Ideally, the preferred combination of the various uses would result in the optimization of sustained aggregate net benefits over time from the ecosystem. In practice, our understanding of complex ecosystems is insufficient to predict all present and future impacts of changes in the uses of different components of the ecosystems. As an essentially non-polluting and non-degrading activity, well managed capture fisheries do not usually subtract from the benefits which other users can derive from inland aquatic ecosystems. On the other hand, maintaining the integrity of these fisheries may place constraints on alternative uses of the aquatic ecosystem such as the generation of energy, sinks for pollutants, and abstractions for irrigation. Frequently, fisheries have been accorded lower priority because of the perception that alternative uses contribute more to society's welfare.

Conservation of the aquatic ecosystem: Conservation of inland aquatic resources should be viewed within the multi-purpose use of river and lake basins. In most inland waters the principle constraints on the system and its living components come from human activities other than fishing.

Government, at all levels from central to local authorities should set up mechanisms to conserve living aquatic resources compatible with the sustainable use of the basin, the aquatic ecosystem and the water for the whole range of economic and social purposes.

User pays principle: Users of the water and the basin should minimize any deleterious effects and contribute to the mitigation of any impacts of their activities and to rehabilitate the systems when the need for their activity has ceased. The OECD definition of the polluter pays principle is as follows:

1. The Polluter-Pays Principle constitutes for [OECD] Member countries a fundamental principle of allocating costs of pollution prevention and control measures introduced by the public authorities in Member countries;
2. The Polluter-Pays Principle, ..., means that the Polluter should bear the expenses of carrying out the measures, as specified in the previous paragraph, to ensure that the environment is in an acceptable state. In other words, the cost of these measures should be reflected in the cost of goods and services which cause pollution in production and/or consumption;
3. Uniform application of this principle, through the adoption of a common basis for Member countries' environmental policies, would encourage the rational use and the better allocation of scarce environmental resources and prevent the appearance of distortions in international trade and investment.

In principle the contribution by those damaging the resource should be total but this is rarely realizable. Government contributions to mitigation should be seen as a subsidy to the industry. Another generally undesirable form of subsidy is to allow the environment either permanently or temporarily, to bear the cost of the abuse, in which case the cost of the degraded state of the environment will be reflected in the lessened value of the aquatic system for society and a delayed bill for the rehabilitation of the system in the future.

Participation of fishers in the policy making process: Fishers or their representatives should participate in the setting of priorities for basin use representing the interests of the fishery and contributing to the mitigation of any undesirable effects of their own activities.

Fisheries management should promote the maintenance of the quality, diversity and availability of fishery resources in sufficient quantities for present and future generations in the context of food security, poverty alleviation and sustainable development. Management measures should not only ensure the conservation of target species but also of species belonging to the same ecosystem or associated with or dependent upon the target species. The nature of inland fishery resources: Fundamental differences exist between fisheries in lakes and reservoirs compared to those in rivers. Lake fisheries tend to be more independent of short term climatic effects, to be based on a relatively small number of target species and to be located in closed systems. Rivers are highly influenced by year-to-year variations in rainfall, the fisheries are based on large numbers of species and the systems are open. Reservoirs cover a range of possibilities intermediate between rivers and lakes. Management strategies should reflect these differences. General impacts of fishery management: It is impossible to operate a fishery without affecting the original equilibrium of the native stock of fishes. Furthermore much of modern management practice deliberately seeks to influence the composition of the fish fauna to correspond to the requirements of society by introducing new species, eliminating those that are considered undesirable and stocking with those seen as valuable.

The types of fishery: Fisheries with simple gears targeting resources with few species and stable hydrological regimes can be assessed using simple, conventional dynamics models. Regulations can therefore be based on traditional criteria such as limitations on mesh size, length at first capture, numbers of fishers, quotas etc. The multi-species, multi-gear fisheries pursued in waters with the unstable hydrological regimes characteristic of many tropical waters are not amenable to such models and require more flexible approaches, usually by limiting access, that adapt to fluctuations in the stock and the environment. States should prevent overfishing and excess fishing capacity and should implement management measures to ensure that fishing effort is commensurate with the productive capacity of the fishery resources and their sustainable utilization. States should take measures to rehabilitate populations as far as possible and when appropriate.

Overfishing: Biologically it is relatively easy to judge the exploitation status of single species fisheries in terms of maximum sustainable or maximum economic yield. This process becomes more difficult in the case of multi-species fisheries where the same level of catch may be sustained over a wide range of effort but where characteristic changes occur in exploited fish assemblages whereby the larger species and individuals are progressively eliminated from the fishery and there is a resulting downward drift in the size of fish caught.

When this happens the fact that more effort is expended to produce the same catch indicates that there is economic overfishing. The extent of rent dissipation would depend on the comparative values of the catch comprising different species of either mostly large or mostly small individuals as well as the cost of redundant fishing effort. In artisanal and subsistence fisheries, the latter may be small because of the low opportunity cost of labour, much of it of a part-time nature, and the inexpensive nature of the fishing inputs. The change in the value of the catch may also be small as consumers shift their preferences in line with the available fish species and size classes.

Therefore, rent dissipation in such circumstances may be limited and, in any case, often costly to avoid because of the difficulty of either defining, allocating and protecting well-specified rights over the fishery resources or to limit fishing effort through administrative means. The most cost-effective management regime may often be to assign rights to groups or communities of fishers which may guarantee that some resource rent can actually be captured. With regard to biological overfishing, this can be determined using conventional models for single species fisheries.

In multi-species situations overfishing can only be determined with respect to assigned target groups. Only when effort reaches such a level that there is an overall decline in catch can the assemblage as a whole be considered to be overfished. However, before reaching this stage, economic overfishing is likely to occur in the form of redundant fishing capacity and decreased net benefits.

Productive capacity: In many inland waters fisheries are enhanced with the express intention of exceeding natural carrying capacity. This may either be done by creating artificial faunas (by introductions, stocking and elimination of unwanted species), that utilize the available food resources more efficiently, through the fertilization of the water or through artificially engineering the environment to correspond better to the needs of the fish.

Developers should apply such methods with caution and initially confine them to enclosed areas where there is little danger of damaging non-target ecosystems. In other areas efforts in support of endangered fish populations should be made to rehabilitate ecosystems damaged by pollution and environmental modification, by physical interventions or by systematic stocking using established protocols.

Conservation and management decisions for fisheries should be based on the best scientific evidence available, also taking into account traditional knowledge of the resources and their habitat, as well as relevant environmental, economic and social factors. States should assign priority to undertake research and data collection in order to improve scientific and technical knowledge of fisheries including their interaction with the ecosystem. In recognizing the transboundary nature of many aquatic ecosystems, States should encourage bilateral and multilateral co-operation in research, as appropriate.

Scientific evidence available: The present general state of biological knowledge of the functioning of inland aquatic systems for fisheries is more than adequate for the formulation of generalized policies for conservation and management. Because of the fragmentary and dispersed nature of inland systems knowledge of individual systems is far from complete, however. Acquiring full information on the numerous water bodies in existence would be prohibitively expensive and limnologists and fishery biologists have elaborated generalized models that are adequate for broad decision making at the capture fisheries level.

As inputs into any system increase or where the impacts of a management intervention external to the fishery need to be established there will be a need to verify generalized models in the case of the individual water body or water course concerned. Knowledge of the social and economic factors regulating the fishery are generally less available because there has been a far shorter tradition of investigation into these concerns.

Drawing on traditional knowledge: One solution to the constraints imposed by the diffuse nature of the resource is to use local and traditional knowledge. Most rivers, reservoirs and lakes are fished and the fishers are well aware of the general biology of the fish species involved, their breeding seasons, migration patterns, responses to fishing methods, etc. Equally traditional systems have been developed in many parts of the world for management of fisheries, integrating local experience and needs of various participant groups.

Unfortunately, the pace of change over the past decades has destroyed many of the traditional management and knowledge systems. Efforts are needed to study and document surviving local management systems and knowledge bases so that they can be improved, adapted and applied to the changing conditions within the individual basin.

Multinational rivers, reservoirs and lakes: Many lake, reservoir and river basins lie within the jurisdiction of more than one country, or within countries more than one administrative unit. In such waters agreements should be reached for common research programmes, standardised systems of reporting, exchange of information and, where possible, common approaches to management. States and subregional and regional fisheries management organizations should apply a precautionary approach widely to conservation, management and exploitation of living aquatic resources in order to protect them and preserve the aquatic environment, taking account of the best scientific evidence available. The absence of adequate scientific information should not be used as a reason for postponing or failing to take measures to conserve target species, associated or dependent species and non-target species and their environment.

Impacts of external influences on the fishery: In inland waters, where the major impacts arise from activities outside the fishery, the precautionary approach should be extended to all developments within the basin. This means that careful impact assessments should be made of non-fishery projects. Unfortunately major damage does not always arise from a single project whose effects can be readily identified but from a series of minor interventions whose cumulative or antagonistic impacts can be enormous. It is thus relatively easy to quantify and compensate for the impact of a single large dam or irrigation project. It is much less easy to quantify and rectify the diffuse pollution caused by widescale agriculture.

Selective and environmentally safe fishing gear and practices should be further developed and applied, to the extent practicable, in order to maintain biodiversity and to conserve the population structure and aquatic ecosystems and protect fish quality. Where proper selective and environmentally safe fishing gear and practices exist, they should be recognized and accorded a priority in establishing conservation and management measures for fisheries. States and users of aquatic ecosystems should minimize waste, catch of non-target species, both fish and non-fish species, and impacts on associated or dependent species.

The nature of fishing gear: Some inland fisheries depend on a relatively small range of mesh selective gears, for example gill nets or seines. In these cases the progressive reduction in mesh size as a reponse to falling catches is the most common problem which can be resolve through clear setting of mesh limitas accompanied by education and extension. The situation is not so clear in many other fisheries, and river fisheries in particular, which use a wide range of fishing gear adapted to the need to capture many different species and life stages of fish during very differing seasons of the year. Some of these gears are judged to be damaging by administrators and fishers and traditional and formal administrative regulations frequently seek to ban such methods. Fishing gears in multi-species systems are usually linked to social structure and it is difficult to ban or suppress certain gears without disturbing the balance of the society. Any decisions of prohibition of gears should therefore be based on sound biological, social and economic advice, taking into account local views, rather than a reaction to hearsay.

The harvesting, handling, processing and distribution of fish and fishery products should be carried out in a manner which will maintain the nutritional value, quality and safety of the products, reduce waste and minimize negative impacts on the environment.

Preservation of fish in inland fisheries: Inland water fisheries are characterised by numerous small craft usually undertaking short journeys from the nearest landing. In other cases the fishers migrate away from their base to remain in temporary camps for considerable periods. Because of the artisanal nature of the fishery it is rare for there to be systematic attempts to preserve the fish on shorter trips and it frequently arrives at the market in poor condition.

On longer trips or in fisheries where there is a valuable product involved ice is used in some areas. Smoking, and sun drying are, however, the commonest approaches although this implies considerable losses in quality and quantity of the product. The extensive use of wood for smoking has been implicated in deforestation, especially of riparian areas. It is a matter of high priority to develop improved systems for the adequate conservation especially in view of the growing lack of wood for smoking in many areas.

All critical fisheries habitats in marine and fresh water ecosystems, such as wetlands, mangroves, reefs, lagoons, nursery and spawning areas, should be protected and rehabilitated as far as possible and where necessary. Particular effort should be made to protect such habitats from destruction, degradation, pollution and other significant impacts resulting from human activities that threaten the health and viability of the fishery resources.

The conservation of the aquatic environment: Responsibility for the protection of aquatic ecosystems usually lies outside the fishery. States should, therefore, plan for the conservation of aquatic environments in the context of their multi-purpose use. Activities such as dam construction for water supply and power, channelization for navigation and flood control, land drainage and wetland reclamation for agricultural and urban use, waste disposal from urban, mining, industrial and agricultural uses, abstractions for agricultural, industrial and urban supply all have a profound impact on the aquatic ecosystem.

Many of these activities are fundamental to the functioning of modern society and are economically of such importance that their limitation in the interests of conservation become hypothetical. All that can be done in many cases is to keep the number of such interventions to a minimum and to limit their impacts. Users of the aquatic system inflicting damage should contribute to the mitigation of the effects of their activities.

Elements for conservation: Certain basic elements are required of an aquatic system so that it can retain its functionality.

a. The maintenance and restoration of longitudinal and lateral connectivity in rivers in the interests of conserving fish migration patterns through removal of transversal (dams) or lateral (levees) obstructions or the provision of fish pass mechanisms.

b. Restoration or maintenance of main channel diversity in rivers including meanders, point bars, bottom structure, vegetation etc.

c. Maintenance or restoration of floodplains and riverine wetlands. This does not have to be continuous along the river but provision should be made for reserves at intervals along the river where normal flood regimes are maintained.

d. Removal and control of all point-sources of pollution including industrial, urban and mining wastes. Control of diffuse pollution particularly of nutrients into lakes and rivers.

e. Control of processes at basin level particularly deforestation, mining operations in rivers and changes in agricultural practice that can lead to massive siltation which can shorten the lives of lakes and reservoirs and destabilize river channels and floodplains.

States should ensure that their fisheries interests, including the need for conservation of the resources, are taken into account in the multiple uses of the coastal zone and are integrated into coastal area management, planning and development.

The articles addressing coastal zone management also apply to river basin management in inland waters. Many of the issues of conservation in rivers and lakes occur in international basins and require negotiation between states to ensure that damaging practices in states upstream are not transmitted downstream to other riparian countries.

Within their respective competences and in accordance with international law, including within the framework of subregional or regional fisheries conservation and management organizations or arrangements, States should ensure compliance with and enforcement of conservation and management measures and establish effective mechanisms, as appropriate, to monitor and control the activities of fishing vessels and fishing support vessels.

The provisions of this article apply particularly to large lakes and to river systems where there are a few large, centralized landings. More generally ensuring compliance with fishery regulations in inland waters is particularly difficult in view of the linear dispersion of the many small fisheries and landings along rivers and the diffuseness of individual fisheries on small and scattered lakes and reservoirs. The cost and practicability of surveillance by a centralised organization become prohibitive and it is usually advocated that these functions be incorporated into co-management agreements so they are undertaken by the fishery communities themselves.

States should, within their respective competences and in accordance with international law, cooperate at subregional, regional and global levels through fisheries management organizations, other international agreements or other arrangements to promote conservation and management, ensure responsible fishing and ensure effective conservation and protection of living aquatic resources throughout their range of distribution, taking into account the need for compatible measures in areas within and beyond national jurisdiction.

International basin authorities: Most lakes, rivers and reservoirs lie within the jurisdiction of one administration. However, many major rivers and some large lakes are international. In such cases States are urged to establish international mechanisms for planning for fisheries management and aquatic conservation in the river or lake basin or to use existing basin regulatory mechanisms established for purposes other than fisheries.

States should, to the extent permitted by national laws and regulations, ensure that decision making processes are transparent and achieve timely solutions to urgent matters. States, in accordance with appropriate procedures, should facilitate consultation and the effective participation of industry, fishworkers, environmental and other interested organizations in decision making with respect to the development of laws and policies related to fisheries management, development, international lending and aid.

Allocation of the inland water resource: Given the multi-purpose nature of inland water resource management consultation and negotiation between the various parties is essential in establishing a coherent resource use policy. Such a policy should centre on the definition of clear allocation of water and the aquatic environment among users as well as a definition of the benefits and responsibilities of each in maintaining the aquatic resource in good health. In actual management regimes one or a few uses of the water tends to assume dominance and other uses have to adjust to its exigencies. This means that the other sectors are in effect involuntarily subsidising the primary use by loss of direct or indirect benefits (opportunity costs).

States should cooperate in order to prevent disputes. All disputes relating to fishing activities and practices should be resolved in a timely, peaceful and co-operative manner, in accordance with applicable international agreements or as may otherwise be agreed between the parties. Pending settlement of a dispute, the States concerned should make every effort to enter into provisional arrangements of a practical nature which should be without prejudice to the final outcome of any dispute settlement procedure.

Disputes: Most disputes in inland water fisheries derive from conflicts between thc fishery and other interests such as farming, flood control and land drainage, navigation, power generation etc. Disputes can also arise among different groups of fishers one of the most fundamental of which is that between recreational and commercial fishers. However, differences also arise between subsistence and professional fishers, between resident and migrant fishers or between groups using different types of gear.

In the past such conflicts have tended to have been resolved within the traditional framework of the fishing communities but as such frameworks have broken down mechanisms for conflict resolution have declined. States can contribute to the resolution of such differences by making clear regulations and decisions as to resource allocation among the various interested parties within the fishery.

In international water, disputes are likely to arise where fish migratory pathways are interrupted, ecological disturbances are transmitted from upstream, breeding sites are destroyed etc. Problems will also be experienced with regard to the cost - benefit distribution of enhanced fisheries where more than one state is involved. In such cases international agreements within the river basin framework are needed to protect all parties.

States, recognising the paramount importance to fishers and fishfarmers of understanding the conservation and management of the fishery resources on which they depend, should promote awareness of responsible fisheries through education and training.

They should ensure that fishers and fishfarmers are involved in the policy formulation and implementation process, also with a view to facilitating the implementation of the Code. Education and training: Training and extension in inland water fisheries is universally important to allow the fishers to participate more fully in the negotiation and decision making processes that are a feature of modern inland fisheries management especially under devolved and co-management regimes. It also becomes more important as the sophistication of management increases with the adoption of enhanced systems where the cost-effectiveness of practices like stocking and fertilization become critical. Where appropriate, States should therefore establish training and extension schemes to assist fishers in making the transition from centrally administered capture fisheries to locally managed enhanced systems.

States should consider aquaculture, including culture-based fisheries, as a means to promote diversification of income and diet. In so doing, States should ensure that resources are used responsibly and adverse impacts on the environment and on local communities are minimized.

The nature of enhancements: A range of practices are deployed in inland waters to raise productivity and in particular production of selected species. These include:

- Introduction of new species to exploit under-utilized parts of the food chain or habitats not colonized by the resident fauna;
- Stocking natural waters to improve recruitment, bias fish assemblage structure to favoured species or maintain productive species that would not breed naturally in the system;
- Fertilization to raise the general level of productivity and overall availability of food for fish thereby increasing growth and survival of the individual fish and increasing their number and biomass;
- Engineering of the environment to improve levels of reproduction, shelter, food resources and vital habitat;
- Elimination of unwanted species that either compete with or predate upon target species;
- Constituting an artificial fauna of selected species to increase the degree of control and the yield from the system;

- Modification of water bodies by cutting off bays and arms or bunding off areas to serve for extensive and intensive fish ponds to increase control and nutrient flows;
- Introduction of cage culture and parallel intensification of effort of the capture fishery;
- Aquaculture through management of the whole system as a fish pond;
- Genetic modification of cultured species to increase growth, production, disease resistance and thermal tolerance of the stocked or cultured material.

The balance between conservation and management: It is clear that enhancement methods of this type are deliberately aimed at changing the productivity of the water and the nature of the fish stock. As such they conflict with the conservation-oriented requirements of the Code. However, these techniques, in common with aquaculture, are aimed at increasing net incomes, as well as, secondarily, the overall supply of fish available for human consumption. As most of them require high levels of inputs of fish seed, feed, fertilizers etc. they can only be regarded as sustainable in the sense that any agricultural activity is sustainable.

That is that the practice may continue from year-to-year at the same levels of input and offtake without noticeable degradation to the natural support system. Furthermore every attempt should be made to isolate the waterbodies which are subject to enhancement so that nutrient-rich effluents or escapes of stocked fish are kept to a minimum.

PRINCIPLES OF DESIGNING INLAND FISH FARMS

1. Site Selection

The site for a fish farm should be selected with great circumspection and care, since the site and general arrangement will control the economics of operation.

Considerations in Site Selection

- Sites should be selected for fish farms only where water of the required volume and quality is available at the times needed for operating the farm.
- Preference should be given to sites where a gravity water supply to the farm is possible.
- The quality of the water available must be such that the desired fish can be raised, e.g. fresh, brackish or salt water.
- Gravity drainage of the ponds should be possible.
- The fish farm should be sited primarily in areas unsuited to other agricultural uses.
- The soil in the area selected should, if possible, be impervious.

- For low construction costs, plain areas with slope less than one percent should be selected.
- The site should be in the vicinity of transportation routes, or where the access road can be constructed economically.
- In the proximity of inhabited areas, considerations of public health and the necessity of guarding against poachers should be kept in mind.
- Fish farms need electrical power, so the possibility and cost of connection should be considered.
- Any existing electric power line must be excluded from the area envisaged for the fish farm.
- Skilled operators are essential for operating major fish farms efficiently. An important consideration in site selection is, therefore, to provide an attractive environment and facilities for both professionals and operators.

Basic Principles of Arrangement

Fish farms consist of many ponds performing different functions in fish production. Their relative positions, as well as their connection to the water supply and drainage facilities, power supply and transport roads, in fact the general arrangement of the fish farm, has a major influence on the operating costs. In deciding on the general arrangement, the following should be considered:

- The farm centre, consisting of operating buildings and living quarters, should have good road access.
- The facilities requiring frequent attendance, such as the hatchery, rearing and nursery ponds, holding ponds and pumping station, should be as near as possible to the centre.
- Separate filling and drainage possibilities should be provided, if possible, for each pond.
- Transport of feed from the grain storage to the ponds and of the fish harvested to the holding ponds should involve short hauling distances.
- The holding ponds should be close to the common external cropping pits serving several ponds.

The foregoing principles can be realized best by an arrangement in which the operating buildings, the holding ponds and the water control structures are in the vicinity of the geometrical centre of the area. Wherever the terrain conditions permit, this type of arrangement should be adopted.

Preparatory Work

The layout of the site is governed by the combined requirements of operation and the particular site conditions. Prior to designing work, all information should be collected regarding technological requirements and the data particular to the site.

Technological Requirements

- Species of fish to be produced.
- Sequence of operations envisaged: production from hatching to market fish, or production of market fish alone.
- Method of fry production.
- Production quantities envisaged.
- Methods and possibilities of nutrient supply to the ponds, such as organic manure, duck farming, fertilizer application.
- Feed distribution (fish feed, grain feed, etc.).
- Transport management - method and means of in-farm and external transportation.
- Buildings required (operation, social-cultural amenities, grain store, equipment shed, repair shop, housing, etc.).

General Technical Data

- Data on existing water uses affected.
- Data affecting water supply and drainage.
- Future development plans for the area.
- Data on other facilities (roads, railways, etc.)
- Property conditions and data.

Geodetical Data

Topographic surveys to the scale 1:500 to 1:5000 and with contour lines of 20-25 cm (perhaps 1 ft) vertical spacing are needed for the entire fish farm area, in order to permit designing complete pond drainage and earthwork volume estimates of the required accuracy.

On any existing earth structures (embankments, canals, etc.) cross sections should be taken at 50-100 m spacing, with points spaced sufficiently close to each other to plot the actual terrain with ± 20 cm accuracy on the cross sections of 1:100 scale.

Cross sections should be plotted at the sites of major structures (intakes, road crossings, etc.)

Cross sections at not more than 500 m intervals and continuous profiles are needed on the connecting stream or canal extending for the distance affected by the fish farm (e.g. to the backwater limit).

The topographic survey should be connected, especially as regards elevations, to the national survey network.

Hydrological and Meteorological Data

Where water is obtained from a natural stream, data must be acquired on the stages and flow rates to be anticipated at the diversion point in the periods of pond filling and for compensation of water losses. Water supply should be designed for a flow rate of 80 percent probability.

In the case of ponds through which floods must be conveyed, or the dikes which are required to retain floods on the stream, the designer will require also data on design flood levels and discharges. The probability of occurrence of the design flood is normally specified by the competent water agency. In the absence of such specification, the flood of 1 percent probability of occurrence (once in a hundred years) should be adopted as the design flood. In the case of minor ponds, where a dam failure would cause no other losses, a flood of say 3 percent probability might be adopted as the design flood. The retention capacity of upstream ponds is taken into account in estimating the design flood.

Data on the peak values of monthly evaporation and rainfall are needed for estimating the water demand.

Data on the monthly average and extreme temperatures are needed for selecting the species of fish for farming, for planning the necessary feed supply rates and for designing the holding and storage facilities of live fish.

The annual volume of sediment entering the ponds should be estimated. Data are needed on the direction and highest speed of wind prevailing in the area in order to design wave protection.

Geotechnical Data

The geotechnical explorations should be extended to the entire area of the fish farm, to provide data on the soil stratification in the pond area, under the dikes, along the canal traces and at the sites of structures.

The data obtained by soil explorations should be suited to estimate:

- Seepage losses
- Underseepage conditions and the hazard of piping failure
- Stability of the dikes
- The required degree of compaction
- The allowable flow velocity in the supply canals, and
- The foundation of the structures.

The methods of exploration and laboratory testing, as well as the interpretation of the results are described in more detail in Chapter 5 on soil characteristics for aquaculture farms.

Water Quality Data

The water supplied to the fish ponds must not contain pollutants and toxic substances detrimental to fish life. The composition of the feed water should be subject to quality analysis, including the following:

- Oxygen content
- PH value
- Total salts content
- Ammonia content
- Free CO2 content
- Phenols, oil and tar content.

The water quality analyses should be such as to enable prediction of the interactions between the soil and the feed water.

PURPOSE AND DIMENSIONING OF FISH FARM STRUCTURES

Hatchery

The most important structure in the technological process of artificial reproduction is the hatching house, where the following operations are carried out:

- Preparation of the breeders prior to stripping.
- Treatment of the breeders (hypothysis treatment, tranquilization, stripping, etc.).
- treatment, fertilization, spawning of the eggs.
- Treatment of the larvae.
- First feeding of the fry.
- Laboratory functions related to artificial reproduction (water analysis, egg analysis, detection of fish pests, chemical and medicament dosage, etc.).
- Packaging of fry.

The equipment needed in the hatchery includes breeder tanks, hypophizing and stripping table, hatching jars, fry holding vessels.

Fry Rearing Ponds and Basins

The small fry are reared from the age of 4-5 days to 3-4 weeks in the fry rearing ponds and basins. These should be arranged in places sheltered from wind, on impervious soil and close to the hatching house and road.

The preferable size of the fry rearing ponds ranges from 100 to 1 000 m2 . Within this range the actual size of the fry rearing pond depends on the number of fry which can be introduced within 1-2 days. The pond size should therefore be adjusted to the capacity of the hatching house.

The water supplied to the fry rearing ponds should be rich in oxygen and must contain no chemical or biological pollution. Water containing much suspended sediment must be allowed to settle first. The supply system should be dimensioned to permit the water in the pond to be exchanged twice daily. In densely populated ponds, a distribution network of perforated pipes should be provided at the bottom for the uniform distribution of inflow water. The fry rearing ponds are lined with stone or concrete.

Circular basins are made with diameters from 4 to 6 m and about 1 m depth. The freshwater is introduced tangentially through nozzle pipes to maintain the water in permanent circulation. The effluent is withdrawn from the centre of the basin.

Rectangular basins should have a ratio of short to long sides from 1:2 to 1:4. The water is introduced to, and withdrawn from, these basins along the short sides. The drainage system of the fry rearing ponds should be dimensioned hydraulically to permit each basin or pond to be drained in four hours. Sluices equipped with box traps should be used to collect the fry. For a guideline to determine the total area of all fry rearing ponds, as a general rule 100 to 200 feeding larvae require 1 m2 of water surface area.

Nursery Ponds

The fry reared in the fry ponds are transferred for further growing into the nursery ponds at the rate of about 100 000 per hectare. The optimal size of the nursery ponds ranges from 1 to 10 ha, but the maximum size should not surpass 30 ha. The water depth should be 1.0 to 1.5 m. The nursery ponds should be arranged in the vicinity of the fry ponds, preferably in a manner to permit direct transfer of the fry to the nursery pond, together with the water from the fry rearing ponds. The water supplied to the nursery ponds should be both chemically and biologically clean. In general no flow through the ponds is necessary, but provision must be made to compensate for any water losses.

Complete and rapid pond drainage is an essential requirement. For this reason the bottom of the ponds should slope towards the outlet and should be so graded as part of the construction work on the ponds. The body weight of around 100 g is attained in these ponds during 160 to 220 days. Fishing should preferably be made possible from external cropping pits.

Production Ponds

These ponds serve to grow yearlings to full grown fish. The ponds are designed according to the following specifications:

The ponds vary from 20 to 100 ha in size, but the largest pond size should be determined by taking the total fish production envisaged into consideration, so that the total fish production in a single pond should not surpass 100 t. For example, for attaining a total unit yield of 2.5 t/ha, the ponds should not be larger than 40 ha.

The ponds should be from 1.2 to 1.8 m deep, with pond depth of 1.8 m especially in regions with elevated temperatures. Water depths less then 0.6 m will result in reduced production.

Filling the ponds should not take more than 2 to 10 days, depending on pond size. No continuous flow through the ponds is required, but the losses due to evaporation and seepage must be compensated for during the growing season. Wherever possible, the production ponds should have filling and drainage systems permitting them to fill and drain independently of the other ponds.

Complete drainage is an essential prerequisition of intensive fish production. In dimensioning the drainage system, an essential criterion is that the number of days required to drain a pond should equal the square root of the pond area in hectares. For example, the draining sluice of a 30 ha growing pond should be dimensioned so as to make possible drainage in or 5.5 days. Another requirement is that the drainage period of the entire fish farm must not surpass twice the square root of the total pond area. For example, the drainage canal of a 500 ha fish farm should be capable of removing within or 45 days the water from all ponds of the fish farm.

The ponds are stocked with fish of 100 g body weight at the rate of 3 000/3 500/ha and the weight gain to 1 500 g/unit is attained in about 300 days.

For advanced fishing, external cropping pits should be provided for the ponds. The feed, manure and fertilizers should be stored in the vicinity of the ponds. For transporting feed, as well as the fish, a sound road network is essential.

Fish Holding Facilities

The fish harvested from the ponds must be stored in holding ponds until they are marketed or processed. The capacity of the holding ponds is governed by the composition of the production according to species, the yield envisaged, the time schedule of fishing and marketing or processing. The breeding stock should be stored, if possible, in the production ponds, rather than in the holding ponds.

For holding purposes small 500 to 2 000 m2 basins or ponds of 0.2 to 10 ha area should be provided. Under climatic conditions similar to those of Hungary, the basins are termed "wintering" ponds or basins. The average depth of the holding ponds should be from 1.8 to 2.2 m. The most important requirement to be observed in connection with the holding ponds is that the dissolved oxygen content of the water in them should never sink below the 2.0 mg/l level. Oxygen can be supplied by introducing a feed water saturated with oxygen, by allowing water to flow through the pond, or by injecting oxygen or air into the holding ponds. Air injection must be accompanied by passing flow through the pond in order to remove the products of metabolism. At high rates of air injection, the rate of flow through can be reduced to around one-third of that without air injection.

To estimate the flow rate through, or air injection rate into the ponds, information is required on the oxygen consumption of the fish stored, the oxygen content of the feed water and the efficiency of the air injection system.

The oxygen consumption rate of fish depends on the water temperature, or on their body temperature which is identical. The rate of oxygen consumption increases with temperature. Conversely, cold water is capable of absorbing more oxygen than warm water. As an example, the hourly oxygen consumption of carp of 1 kg average body weight is;

10 cm3 at 5 C

25 cm3 at 10 C

50 cm3 at 50 C

Each litre of saturated water contains the following amounts of oxygen:

8.91 cm3 at 5 C

7.87 cm at 10 C

7.04 cm3 at 15 C

When estimating the necessary flow rate, it should be remembered that the fish do not consume the entire oxygen content of the water for breathing, so that the residual oxygen content of the effluent from the ponds is still around 1 cm3/l.

In the absence of more accurate computations, the flow rate through the holding ponds can be estimated at 1.0 l/sec for each ton of fish stored.

The allowable fish density in the holding ponds ranges from 5 to 15 kg/m2 pond area.

FISH POND ARRAGEMENTS

The arrangement of the fish ponds at any particular fish farm site is normally dictated by topographic conditions. The following types of pond arrangement are usual:

- Barrage ponds.
- Contour ponds.
- Paddy ponds.

The first two types of pond system are typical in rolling terrain, while paddy ponds are adopted in flat land. These types of arrangement differ from each other in their method of construction, structures and operation.

Barrage Ponds

Such ponds are established in a valley of flat to medium longitudinal slope, by closing the valley with a low dam at a suitable site. A site is considered suitable if the shortest possible dam will create the largest possible pond. Thus, dams can be constructed most economically at a narrow point in the valley situated at the downstream end of a flat, wide section. The floods on the stream will pass through the ponds and must be released without causing damage to the dams, dikes and structures, and without allowing the fish to escape. For this purpose a spillway must be provided.

Contour Ponds

The ponds established behind a longitudinal dike in the valley will not be affected by floods, these being passed along one of the valley sides. In other words, fish ponds are created on one side of the valley alone. To create gravitational supply to the ponds, the stream is blocked by a weir, the water being diverted through a gated intake to the supply canal. From the latter, each pond can be filled separately, while separate drainage is possible to the original stream bed except during flood conditions. The relocated stream bed must be dimensioned to carry the design flood safely, while the crest of the longitudinal dike running parallel to the stream must be raised above the design flood level. This arrangement is only possible on wide valley floors.

Paddy Ponds

Each pond of fish farms in flat plain areas is surrounded by a dike. The flat terrain offers wider opportunities for the favourable arrangement of ponds, in which each pond can be filled and drained independently, thus creating the possibility of intensive fish farming.

Earth Structures

Earth structures form an important part of any fish farm project. The dams and dikes are constructed of earth, while the supply and drainage canals, the seepage drain ditches, the pond drains and the harvesting pits are excavated in earth.

Dams and Dikes

The dams and dikes are normally constructed of the soil material available at the site. The dimensions and cross sectional shape of the dikes are governed by the purpose they are intended to serve and the material available for construction. The main principles to be observed are:

- The crest width should not be less than 2 m.
- The dike crests serving as unpaved roadways for internal transportation should be at least 3 m wide.
- Paved roads on the dike crests should be made with at least 1 m wide unpaved shoulders.
- The centre line of the dike crest should be elevated by 15-20 cm relative to the shoulders.
- The crest elevation above the highest water level, the safety freeboard, is controlled by the maximum wave height anticipated. In the absence of calculations, a freeboard of 50 cm should be provided.
- The slope inclinations should be selected as outlined in the notes on soils engineering.
- Where a reed belt is envisaged for wave action control either a horizontal 3 m wide reed berm should be envisaged with the surface not more than 0.5 m below the normal pond level, or a slope inclined at 1:4 to 1:8 should be formed starting 0.3 m below the normal water level.

Feeder Canals

The feeder, or supply canals , carry water from the intake to the individual ponds, and are used occasionally as waterways for barges.

The feeder canals should be dimensioned to allow the various types of ponds to be filled within the times mentioned. Moreover, they should enable all ponds of the fish farm to be filled in less than 50 days.

The longitudinal profile of the feeder canal should be designed to ensure a canal water level at least 0.1 m higher than the normal water level at the intakes to the individual ponds. In estimating the water level in the feeder canal, the backwater caused by the structures should normally be taken into consideration.

Depending on the seepage losses anticipated from the feeder canal, some kind of lining, or other method of seepage control, may be necessary.

The feeder canals are made with trapezoidal, or mixed, cross-sections. The bottom width and water depth are found from the hydraulic calculations. The dikes along the feeder canals should have at least 2.0 m wide crests. The slope inclinations depend, just as in the case of the pond dikes, on the quality of soils, but should not be steeper than 1:1.5. The safety freeboard along the feeder canals should normally be 0.5 m, but can be reduced sometimes to 0.3 m over the terminal sections.

Canals to be operated at freezing temperatures should be designed for water depths of more than 1.0 m.

Drainage canals

The drainage canals serve to convey the water from the ponds to the recipient stream when the ponds are drained.

The lowest water level in the canal at the outlet of the drain sluices should be at least 20 cm deeper than the lowest bottom elevation in the pond, or than the bottom of the harvesting pit. Where this is impossible to achieve, pumps should be considered to ensure complete drainage of the pond.

The conveying capacity of the drainage canals should be estimated starting from the drainage times and periods mentioned for the various types of pond.

Drain Ditch

These ditches serve to drain any seepage from the ponds and the feeder canals, and the external runoff that is prevented by the pond dikes from draining towards the recipient stream and, thus, to protect the surrounding areas against the potential damages caused by such waters.

The bottom of the ditch should nowhere be less than 0.3 m below the terrain level and the water should not be allowed to overflow, even when carrying the design discharge.

The drain ditch follows largely the trace of the extreme dike. The distance from the dike depends on the soil conditions and stratification (see notes on soils engineering, "piping control"). The distance between the outer toe of the dike slope and the edge of the drain ditch should not be less than 2.0 m.

Internal Pond Drains

The water from undrained depressions within the pond area is conveyed to the outlet sluice, from where the fish are carried to the cropping pit. The slope of these drains should be steeper than 0.1 percent, while they should be cut deeper than 0.2 m. The minimal bottom width desirable is 3.0 m and the sides should be sloped at 1:3, or flatter. The earth excavated from these drains might possibly be used for constructing the dikes or for filling any depressions. Any material that cannot be used for such purposes should be deposited at a distance of not less than 5 m from the drain, so as not to interfere with draining the pond, and to prevent it from sliding back into the drains.

Borrow Pits

Where it is impossible to secure the earth needed for constructing the dikes from the cuts envisaged, separate borrow pits must be opened. Material may be obtained from areas outside the ponds, but also from within the pond as part of the bottom levelling work, or as special borrow pits. Attempts should be made to obtain the dike material from levelling the pond bottom, which should slope towards the outlet sluice. The bottom should be levelled to an accuracy of ± 15 cm. Any internal borrow pit should be designed by observing the principles outlined in connection with the pond drain network.

Internal Harvesting Pits

For harvesting, cropping pits should be placed in the vicinity of the outlet sluice, either within or outside the pond. An internal pit serves one pond only, while two or more ponds can be connected to an external pit.

Internal harvesting pits may be arranged parallel or perpendicular to the dike. The bottom area required for the cropping pit is around 40 m2/ha. To accommodate the length of the nets in current use, the pits should be 10 to 25 m wide. The differential elevation between the two ends of the bottom of the pit should be 20 cm. The preferable depth of the pits ranges from 0.6 to 1.0 m. The internal borrow pits and pond drains should be connected to the harvesting pit, so that their bottom at the entrance should be at least 20 cm above the pit bottom.

External harvesting pits are connected to the pond outlet sluices over drainage canals of varying length. Fresh water should always be introduced into the external harvesting pits during the period of harvesting. External harvesting pits need not be dimensioned for the entire fish population in the ponds, since the fish are admitted intermittently to the pit. The preferable area in plan varies from 15 50 to 20 70 m. The bottom of the external harvesting pits should be at least 30 cm deeper than the deepest point within the pond and an additional differential elevation of 20 cm is necessary between the two ends of the harvesting pit. The crest of the dike around the external harvesting pit should be designed so as to have a pit depth of 0.8 to 1.0 m below it.

The slopes of the harvesting pit should not be steeper than 1:2, while over the section where the net is drawn out they should be inclined at 1:4.

DIKE PROTECTION

The dikes require effective protection against the destroying effects of waves. Even the surfaces of earth structures not exposed directly to pond water may often require protection against rain and wind erosion, depending on the prevailing climatic conditions.

Wave Action

The length and height of wind generated waves depend on wind speed, the fetch length and the depth of water. Of the large number of empirical wave height formulae, the one suggested by Dyakova is

h = 0.0186 W0.71 B0.24 H0.54

where h is the wave height in metres, W the wind speed in m/sec, B the fetch length in km and H the water depth in metres.

Biological Protection

Biological protection consists of establishing a vegetation cover as a means of controlling both wave action and erosion, provided that such vegetation can be planted and grown under the particular conditions, and is compatible with the requirements of fish production.

Information on the plants potentially applicable and their biological requirements in any particular area should best be obtained at the site. Under the conditions prevailing in Europe, a reed belt has been found most effective for controlling wave action.

For effective wave action control, the reed belt should be 4 m wide for the first 1 000 m length of fetch, and should be increased by 2 m for every subsequent 1 000 m increment of fetch length. The density of the stand should be at least 70 reeds/m2 and each reed should have a diameter of not less than 8 mm above the water level.

The wave control reed belt is planted on the reed berm of the dike. Over the surface on which the reed is to be planted a topsoil or humus cover of at least 20 cm thickness should be spread, in which the reed is planted either by seeding, as a soil root mixture or as root or shoot cuttings. Except for propagation as shoot cuttings, reed planting operations are confined to the winter hibernation interval. The shoots should be planted at the time when the fresh shoots have developed 7 to 8 leaves.

Whichever method of reed propagation is adopted, the water demand of the growing plants must be satisfied. Without the possibility of irrigation, the result of reed planting will be questionable. In the first period of growth, up to the appearance of 3 to 4 leaves, the soil should be saturated, but a permanent water cover is definitely undesirable. On the slope surfaces and dike crests exposed to rain and wind erosion, a sound grass cover is often the most economical method of control. For grass planting the surfaces should be covered by at least 10 cm of topsoil, but this can occasionally be replaced by manure and fertilizers. The grass cover is established on properly prepared surfaces most expediently by seeding. The desirable seed mixture should be determined with due consideration to the soil type and climatic conditions. The grass cover must be mowed regularly and in dry climates provision must be made also for irrigation. Where irrigation is impracticable, the costs of grass seeding will be wasted.

Wave Control Linings

Where no biological protection is possible, the earth structures exposed to wave action must be protected by other types of lining. Guidance to designing such linings can be found in the notes on soils engineering.

STRUCTURES

The structures normally used for supplying water to, and draining water from a fish farm are of types familiar in hydraulic engineering. Those particular to fish farm projects will be dealt with here.

Monk Sluices

Monk sluices are the structures generally used for supplying water to, and draining water from, fish ponds. By means of stop-log closures and fish screens or racks, the operator is free to decide whether the lower or the upper water layer from a pond should be released.

Monk sluices consist of two main parts, namely the shaft and the horizontal pipe. Three pairs of slots are incorporated in the shaft, one for the fish screen, the other two for the stop-log closure. The shaft is normally on the upstream side, thus inlet monks for filling have their shaft on the side of the supply canal, while outlet monks are on the pond side. The pipe crosses under the dike and serves also to dissipate the energy of the water falling over the stop-logs. The pipe may be built with circular, horseshoe, or rectangular cross sections. For higher conveying capacity twin pipes may be provided.

Access to the shaft is made possible by a walkway at least 0.5 m wide. For working on top of the shaft, a platform is cantilevered therefrom. A handrail must be provided along the walkway and the platform to prevent accidents.

Open Sluices

Open sluices are built where the passage of barges is required, or where the discharges to be passed are higher than those which normal monk sluices are capable of carrying. Their application is also needed for the inlet sluices of external cropping pits, to facilitate the passage of fish. They are preferable also in the screen-box structure of nursery ponds where drainage or fishing of two nursery ponds can be handled by means of a single structure.

In the longitudinal section the structure follows the cross-sectional shape of the dike. The open cross section is U-shaped. To enable traffic to pass over the structure, the open U can be bridged by a slab.

With closing gates provided on both the upstream and the downstream ends, such open sluices can be designed as locks.

The closing mechanism in open sluices may be stop-logs or vertical-lift steel gates. Openings wider than 1.5 m are cumbersome to close by means of stop-logs and here steel gates with proper hoisting gear are the better solution. The hoisting gear of frequently operated gates should be driven by an electric motor where power connection is possible. The steel gates should be of the double-leaf type which can be released from either below or above the gate, as required by the operator.

Spillways

Spillways are essential components for the safe operation of fish ponds formed by valley closure. These serve to pass the floods arriving at the ponds without damage, while preventing the water level in the ponds from sinking below the normal operating elevation and the fish from escaping from the pond system. The spillways should be dimensioned to pass the design flood without damage. The structure must be capable of passing at least the 10 percent probability flood without any human interference. At higher flood discharges human control may be necessary at the closing mechanisms of the structure.

A wide variety of spillway designs have been developed, of which a few simple types will be described. Simple unlined spillways are earth canals arranged at the junction of the dam and the valley side and having a sill level corresponding to the normal operating water level in the pond. Normally, no fish screen is provided. For stabilizing the sill, a recessed beam of concrete or hand-placed rock may be advisable. The depth of flow over the spillway, or the highest allowable flow velocity must be determined with due regard to the local soil conditions so as to prevent scouring. The surface of the spillway may be protected against scouring by a grass cover, reinforced if necessary by rows of buried brushwork or hand-placed stone ribs.

Improved unlined spillways are similar in design to the foregoing but with fish screens between piers and lined for some distance upstream and downstream of the screen. The piers have cantilevers at the top, on which a service walkway is supported to permit cleaning the screen.

Reinforced concrete spillways with recessed sill have the sill elevation below the normal operating water level in the pond, which is maintained by means of stop-logs. The fish are prevented from escaping by a fish screen. To pass major floods, the stop-logs must be lifted clear of the water. For ease of handling the stop-logs, the spans between the piers should not be longer than 1.5 m. The sill should not be lower than 0.6 m below the normal pond level. The piers should carry a walkway or a footbridge.

Drop shaft spillways are normally combined with the outlet sluice. Where the subsoil offers good foundation conditions, such structures are capable of passing safely minor flood discharges. Behind the shaft of the monk sluice a drop shaft having a crest elevation at the normal pond level is also provided. The flood discharge is carried by the pipe of the sluice, so that the pipe must be dimensioned for the design flood flow. At the downstream end of the pipe, an energy dissipator is needed. Cleaning of the fish screen around the crest of the drop shaft can be accomplished from the access bridge of the structure.

Fish Control Structures

At valley-dam fish ponds supplied by a perennial stream, measures must be taken to prevent fish from escaping from the upstream pond. A fish screen or an electrical fish deterrent are possible solutions.

The fish screen should extend at least up to the level if the 10 percent flood. The screen must be cleaned at the time of floods, so that a working platform must be provided for the operator. The mesh width of the screen or the spacing of the rack bars depend on the size of fish kept in the pond. In production ponds a bar spacing of 2 cm is sufficient.

INLAND CAPTURE FISHERIES

This chapter deals with inland capture fisheries for food production. Although increasingly important in many areas, there is no systematic reporting of recreational fisheries to the FAO or to any other organization with global scope.

The objectives of this chapter are:

- To update and expand the scope of the 1999 version of the REVIEW OF THE STATE OF WORLD FISHERY RESOURCES: INLAND FISHERIES (FAO, 1999);
- To place inland capture fisheries in its proper context in relation to world fish production;
- To review the status and trends of inland fisheries production at global, continental and country levels;
- To call attention to the importance of inland capture fisheries with respect to food security and human development;
- To examine environmental sustainability in the context of the aquatic environments in which fishery resources reside and fisheries operate; and
- To call attention to the main issues facing inland fisheries.

The present version differs from earlier versions of Circular 942 in the following ways:

- The place of inland capture fisheries in the global production of aquatic food products is expanded to highlight countries where inland capture fisheries production is important relative to other aquatic production sectors.
- Updating of FAO statistical information is through 2001; recent analysis by FAO allows trends to be viewed from 1950; the data have been presented in environmental and capture fisheries-culture combinations.
- Colour maps have been added to illustrate better the country level geographic distribution of important characteristics of inland fisheries, and charts are in colour for better discrimination of comparisons.
- Indicators of the importance of inland capture fisheries production have been added including:
    - Per capita production.
    - Production per unit area.
    - Fifteen-year trends.
    - Environmental Sustainability Index.
    - Human Development Index.

The place of inland fisheries in world fish production - inland capture fisheries relative to marine capture fisheries and aquaculture

This section reviews trends and status of inland capture fisheries production in relation to aquatic production from marine capture fisheries and aquaculture.

Evolution of capture fisheries and culture in global time series

World total inland capture fisheries production reported to FAO reached 8.7 million tonnes in 2001, down slightly from 8.8 million tonnes in 2000. It is important to note at the outset that actual catches overall may be at least twice the quantities reported to FAO, and for some countries underreporting could be even greater (FAO, 1999).

Inland capture fisheries accounted for 6.1 percent of the total (capture fisheries and culture) global production. This compares with 59.8 percent (85 million tonnes) from marine capture fisheries of which about 31 million tonnes were used for reduction to meal and other non-food purposes and 34.1 percent (48.4 million tonnes) from aquaculture, including plants and animals. Long-term trends, from 1950 to the present, show increases for all three categories; however, production from marine capture fisheries has leveled off in recent years while aquaculture has rapidly increased. In contrast, production from inland capture fisheries has exhibited a modest, but steady increase (Figure 6.1).

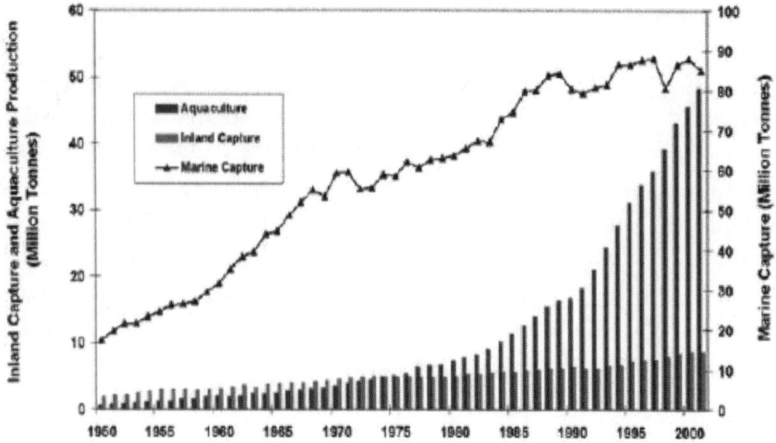

Figure 6.1: Inland Capture Fisheries and Aquaculture with Marine Capture Fisheries 1950-2001

Trends in global capture fisheries and culture by environment

Over the last half century the relatively modest growth of inland and marine capture fisheries is evident with overall increases approaching five fold in that period. This contrasts with the rapid growth of aquaculture in recent times; freshwater culture, mariculture, and brackishwater culture have increased nine fold, six fold, and five fold, respectively, since 1980 (Figure 6.2).

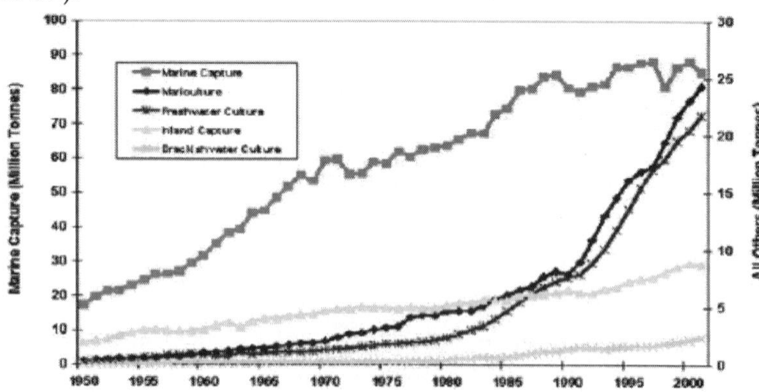

Figure 6.2: Evolution of capture fisheries and aquaculture production by environment 1950-2001

Considering fish production from all sources in the inland environment calls attention to the increasing use of inland waters for culture as well as for capture fisheries. This is evident in the 52-year trends for inland capture fisheries, freshwater culture and brackishwater culture (Figure 6.3).

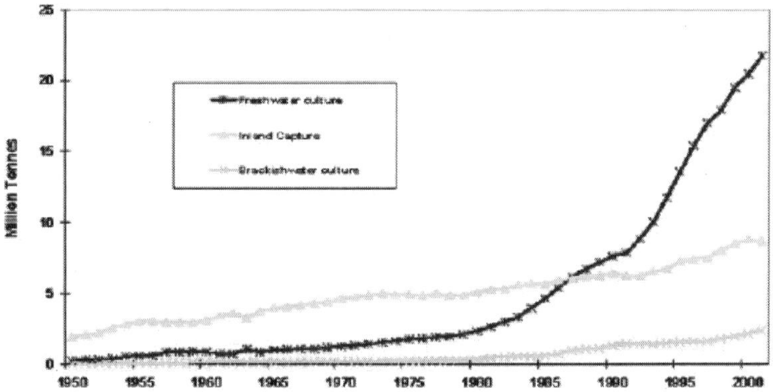

Figure 6.3: Production from inland waters - evolution of inland capture fisheries, freshwater and brackish-water culture 1950-2001

This is particularly true where brackishwater and freshwater culture are extensive. At the level of small individual water bodies, and at the level of parts of larger water bodies, it is well known that there is competition between inland fisheries and aquaculture for space and for access with social and economic consequences.

Global view of the importance of inland fisheries relative to marine capture fisheries and aquaculture in 2001

Two hundred twenty five countries reported fish production in 2001, of which, 150 reported inland capture fisheries production. This compares with 187 countries reporting marine capture fisheries production, and 164 countries reporting aquaculture production.

Among the 150 countries reporting inland capture fisheries statistics there were seven in which inland fisheries provided the only source of fish, and 20 additional countries in which inland capture fisheries could be considered extremely important, accounting for 81 to 99 percent of total fish production from all sources within the country. There were four additional countries in which inland capture fisheries could be said to be very important, accounting for 61 to 80 percent of total aquatic production. These 31 countries together produced 1.8 million tonnes of inland fish equivalent to 20 percent of total inland capture fisheries production.

Global view of the role of China in fisheries and aquaculture

China dominates among the major categories of global aquatic production (Figure 6.4). Therefore, China is considered separately herein.

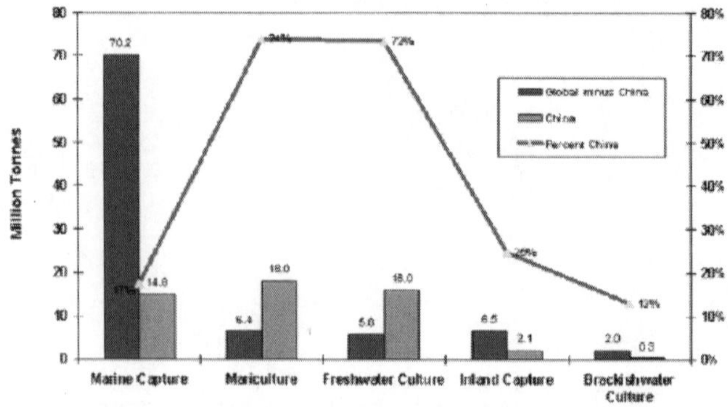

Figure 6.4: China in world capture fisheries and culture production in 2001

The situation is particularly marked in the case of global mariculture and freshwater aquaculture for which China accounted for 74 percent and 73 percent, respectively, in 2001. In comparison, China accounted for 17 percent of marine capture fisheries production, 25 percent of inland capture fisheries production, and 13 percent of brackishwater culture in 2001 (Figure 6.4).

Continental view of capture fisheries and culture production by environment in 2001

Asia dominates production both from capture fisheries and culture viewpoints (Figure 6.5). Inland capture fisheries production from Asia was 5.8 million tonnes; Africa ranked second with 2.1 million tonnes. Inland capture fisheries production from Europe and South America was 0.3 million tonnes, each, and in North America 0.2 million tonnes. Inland capture fisheries from Oceania produced 22 thousand tonnes in 2001.

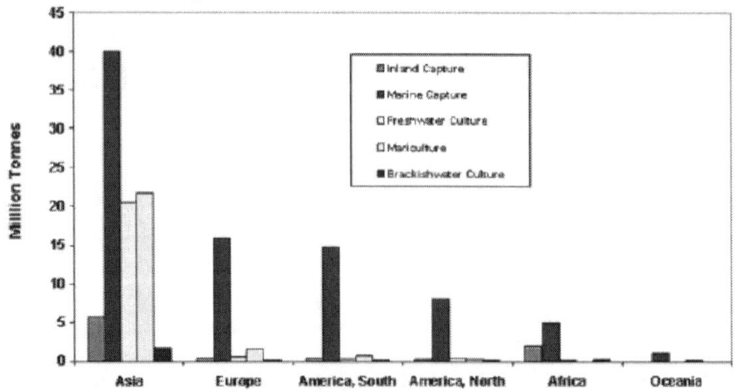

Figure 6.5: Capture fisheries and culture production in 2001 by continent and environment

In Asia, China dominates fish production accounting for 83 percent of mariculture, 78 percent of freshwater culture, 37 percent each of marine and inland capture fisheries, and 17 percent of brackishwater culture (Figure 6.6).

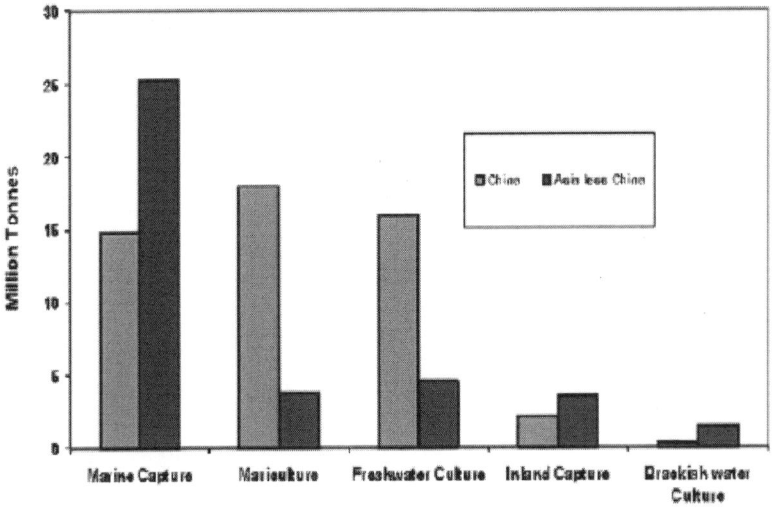

Figure 6.6: The place of China in fish production in Asia in 2001

Fish production from inland waters - inland capture fisheries and freshwater and brackishwater aquaculture production

In inland waters fish and other other aquatic animal production can come from capture fisheries, or from brackishwater or freshwater aquaculture. It is important to partition production from inland capture fisheries from production from inland aquaculture because the former is potentially the most immediate source of high quality protein for the populace. Much of inland capture fisheries production is available for consumption domestically whereas at least part of inland aquaculture production is for export, or too expensive to be available to the low-income stratum of the population.

There were 32 countries among the 150 inland fisheries countries with production from inland capture fisheries as the sole source of fish from inland waters. Additionally, there were 34 other countries in which inland capture fisheries were extremely important, accounting for at least 81 to 99 percent of the freshwater and brackishwater production from all sources, and an additional ten countries where inland capture fisheries were very important, accounting for at least 61 to 80 percent of total inland fish production. These 76 countries produced 3.0 million tonnes in 2001, equivalent to 35 percent of the total inland capture fisheries output.

COMPARATIVE EVALUATION OF INLAND FISHERIES

In this section the importance of inland capture fisheries at the country level is considered from several viewpoints:

1. Production per se.
2. Production per capita.
3. Production per unit area.

Importance based on inland capture fisheries production

In 2001, inland capture fisheries production reached 2.1 million tonnes in China, approximately one million tonnes in India, and 0.7 million tonnes in Bangladesh. The remaining countries among the top 20 produced from 360 000 tonnes down to 92 000 tonnes (Figure 6.6). The top 20 countries account for 84 percent of the total global inland capture fisheries production. Most of the important inland fisheries countries are in Asia (10) and Africa (7).

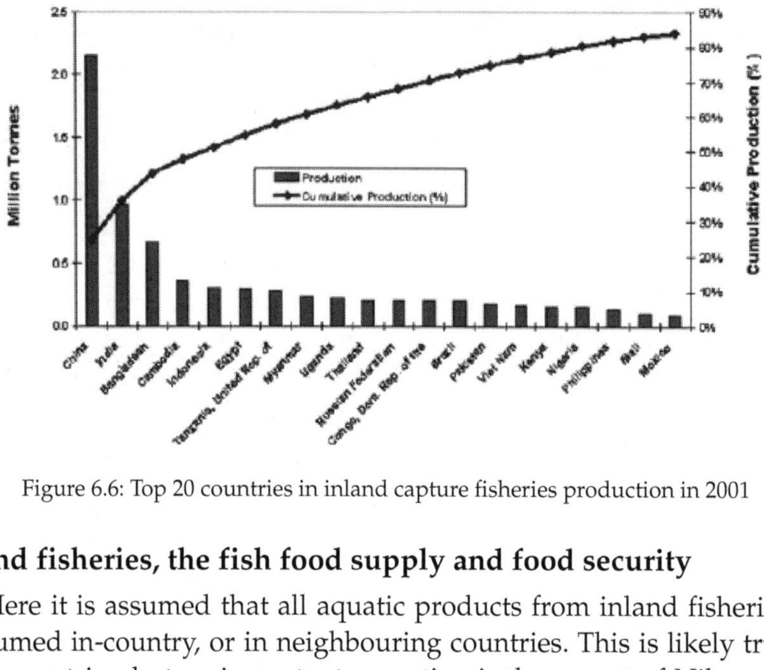

Figure 6.6: Top 20 countries in inland capture fisheries production in 2001

Inland fisheries, the fish food supply and food security

Here it is assumed that all aquatic products from inland fisheries are consumed in-country, or in neighbouring countries. This is likely true for most countries, but an important exception is the export of Nile perch to Europe from Lake Victoria in Africa. Nevertheless, inland capture fisheries production divided by population is taken as an indication of the fish per capita supply from inland capture fisheries.

Cambodia is the preeminent country in per capita inland capture fisheries production with 28.2 kg, exceeding that of the second ranking country, Chad, by a factor of more than three (Figure 6.7). Apart from Cambodia, per capita inland production ranges from 9.7 to 3.3 kg among the other 19 countries in the top 20.

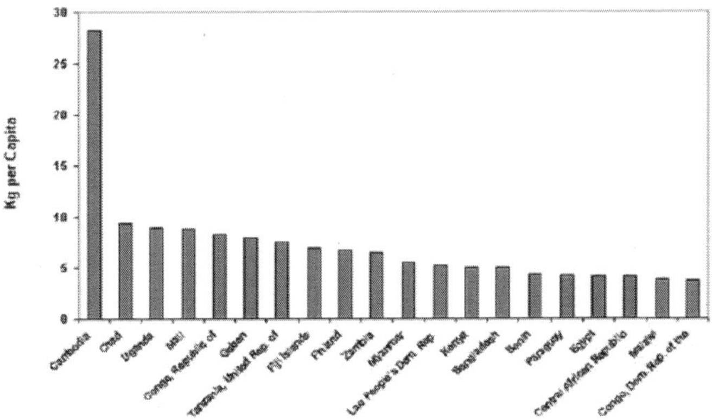

Figure 6.7: Top 20 countries in
inland capture fisheries production per capita

As a comparison with the weighted mean per capita inland capture fisheries production of 3.9 kg, the per capita fish supply from all sources in 2001 was 13.1 kg (excluding China), and the overall fish supply from aquaculture was 2.4 kg. However, in the low-income food-deficit countries, 71 of which are countries reporting inland fisheries production, the per capita fish supply is 8.3 kg, excluding China. Globally, most inland capture fisheries countries have per capita inland capture fisheries values of less than 2.5 kg per person.

Comparing per capita inland capture fisheries production (Figure 6.8) with production per se (Figure 6.8) reveals that there are nine countries that are included in the top 20 in both categories: in Asia-Bangladesh, Cambodia, and Myanmar, in Africa-Egypt, the United Republic of Tanzania, Uganda, Democratic Republic of the Congo, Kenya and Mali. Considering production per capita shifts the focus of continental importance from Asia to Africa. Among the top 20 countries production-wise, ten countries were in Asia and seven in Africa. In contrast, the top 20 in per capita inland capture fisheries include 13 from Africa and only four from Asia. Thus, the importance of both production in its own right, and production per capita is emphasized for the nine countries in common among the top 20, and for the others, emphasis is shifted from Asia to Africa in terms of the apparent important contribution of inland capture fisheries to the national fish food supply. Inland capture fisheries are particularly important in African land locked countries because of the relatively low levels of aquaculture production there.

Importance of inland capture fisheries production relative to land area

Inland capture fisheries production per unit area of a country (tonnes/ km2) is an alternative to production alone as an indication of the importance of inland fisheries in a country. For example, Bangladesh (6.7 tonnes/

km2), Cambodia (2.0 tonnes/km2) and Uganda (0.9 tonnes/km2) stand out among the top 20 countries in inland capture fisheries production per unit area while the remaining 17 range from 0.5 to 0.2 tonnes/km2 (Fig. 6.8). Production per unit of area brings out the importance of inland fisheries in relatively small countries such as Fiji, Benin, Burundi and Rwanda (Figure 6.8). Abetter measure of the importance of inland capture production relative to area would be production per unit of water surface area. However, water surface data are not comprehensively and comparatively available for many inland fisheries countries.

INLAND FISHERIES IN LOW-INCOME FOOD-DEFICIT COUNTRIES

The classification of a country as low-income food-deficit (LIFDC), used for analytical purposes by FAO, is traditionally determined by four criteria. First, a country should have a per capita income below the "historical" ceiling used by the World Bank to determine eligibility for International Development Association (IDA) assistance and for 20-year International Bank for Reconstruction and Development (IBRD) terms. The second criterion is based on the net food trade position of a country (i.e. gross imports less gross exports) averaged over the preceding three years.

Thirdly the self-exclusion criterion is applied when countries that meet the above two criteria specifically request to be excluded from the LIFDC category. In order to avoid countries changing their LIFDC status too frequently - typically reflecting short-term, exogenous shocks - a "persistence of position" is considered which would postpone the "exit" of a LIFDC from the list, despite the country not meeting the LIFDC income criterion or the food-deficit criterion, until the change in its status is verified for three consecutive years.

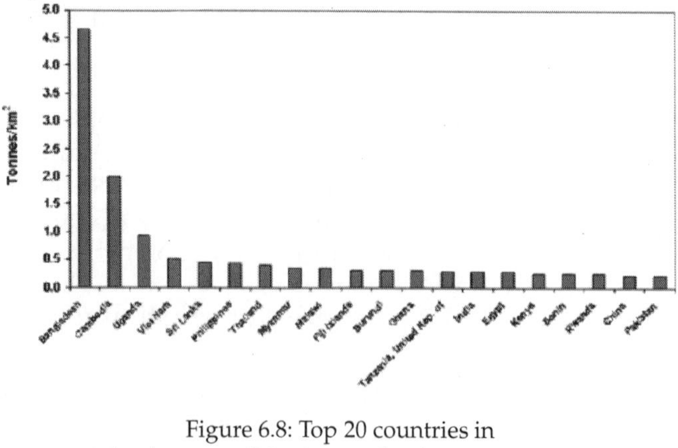

Figure 6.8: Top 20 countries in inland capture fisheries production per unit area

Of the 150 inland capture fisheries countries there are 71 that are LIFD countries. The LIFD countries produce 80 percent, nearly 7 million tonnes, of the world total inland capture fisheries output. These totals are heavily influenced by China. China alone as an LIFD country accounts for 31 percent (2.1 million tonnes) of the total LIFDC inland capture fisheries production. All but four of the top twenty countries in inland capture fisheries production and in inland production per capita in 2001 are LIFD countries.

Among 27 of the LIFD countries inland capture fisheries are the sole source of fish, and in an additional 22 countries inland capture fisheries account for at least 81 percent of the total inland fish production. In an additional three countries inland capture fisheries make up at least 61 percent of inland production. In LIFD countries the contribution of fish to total animal protein intake is nearly 20 percent. Therefore, sustaining inland capture fisheries production in most of the LIFD countries should be an important food security goal.

STATUS AND TRENDS OF INLAND CAPTURE FISHERIES PRODUCTION

Trends in inland capture fisheries production at country level

Although 52-year trends are available for all inland fisheries countries, trends for the most recent 15-year period have been used. Interest should be focused on recent performance, and a 15-year period is sufficient to capture positive and negative effects of varying water quantities on production in terms of lake and river volumes, and in terms of varying inundation of floodplains and other fished shallow water surfaces. It also is a period sufficiently long in which the effects of varying fishing pressure on resources would become evident. Thus, the 15-year trends are an indication of the status of resources as affected by environment and fishing together as revealed by annual production data.

Considering the 15-year (1986-2001) trends in inland capture fisheries production grouped by the numbers of inland capture fisheries countries in each trend category, it can be seen that about half are relatively stable in capture production, about 17 percent are slowly increasing, and about 20 percent are slowly decreasing in production (Figure 6.9).

Relatively few countries are moderately or rapidly increasing in inland capture fisheries, and no countries are rapidly decreasing.

Thus, in terms of numbers of countries in each trend category, the global picture is a relatively positive one with 109 of the 150 countries maintaining stable or better outputs.

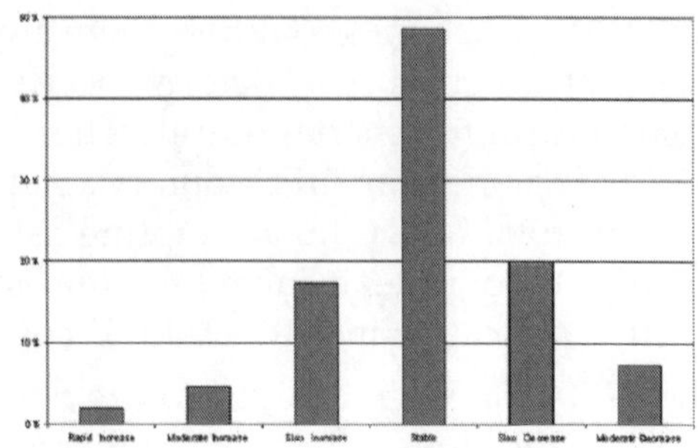

Figure 6.9: Trends in inland capture fisheries production (1987-2001): percent of 150 countries

TRENDS IN INLAND CAPTURE FISHERIES PRODUCTION 1986-2001

Considering the 15-year trends, this time grouped by the quantities of production contributed by the countries in each trend category, it can be seen that countries with stable trends contribute about 35 percent of the total inland production while countries with slowly increasing production contribute about 56 percent of the total (Figure 6.10). Little of the production is in countries where production is moderately or rapidly increasing or decreasing.

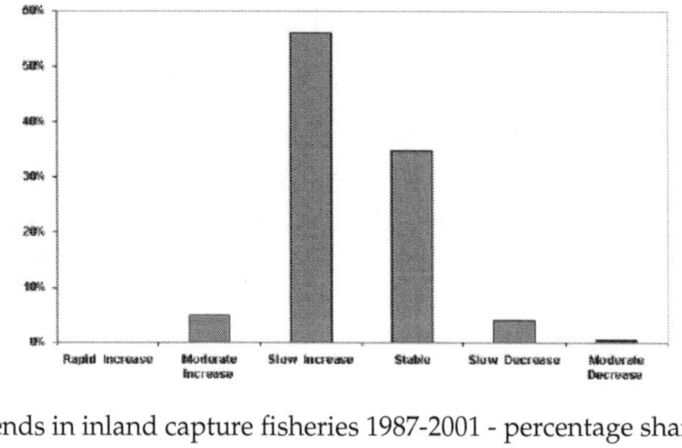

Figure 6.10: Trends in inland capture fisheries 1987-2001 - percentage share of production in 2001

Thus, in terms of the share of production in each trend category, the picture is positive with 81 percent of the production falling into categories where capture is stable or slowly increasing. Identifying the catch - status of the identification of organisms from inland capture fisheries in 2000, and trends from 1980 to 2000 Identification of the organisms in the catch to species level is valuable for the management of fisheries as well as for fishery resource and economic assessments. Not only is the information of use internally in a country, but also for comparative purposes among countries that share river and lake basins, particularly to monitor the status of migratory fishes, other shared resources, and introduced and invasive species.

The objective of this section is to assess the extent to which countries with inland fisheries identify and report organisms in the catch. The assessment is in two parts: the situation in 2000, and trends in reporting from 1980 to 2000 in five year increments. Although data are available for 2001, it was decided to use 2000 as the most recent year so as to avoid bias from countries that report inland capture late. The approach was to calculate an "organism identification score" for each country for each selected year. The score is based on two criteria: (1) the level of detail of reporting, and (2) the importance of the organisms reported at a given level as indicated by the percentage of the total tonnage caught.

Taxonomic coding of the information provided to FAO by each country allows five levels of reporting to be identified: Level 5: Species; Level 4: Genus; Level 3: Family; Level 2: Order, or other taxonomic level (better identified, e.g. Cypriniformes); and Level 1: Order, or other taxonomic level (less well identified, e.g., Miscellaneous fishes). For this evaluation, a point system was used that "rewards" reporting at the most detailed level. For example, five points were assigned to Level 5 for reporting at the species level and one point assigned for reporting at the least detailed level of identification, Level1.

STATUS 2000

There were 149 countries with inland fisheries production in 2000 for which the organism identification scores were calculated. About one-third of the countries scored 20, the lowest possible score, indicating that they reported their inland production only at the least detailed level of identification. There were an additional 23 countries with relatively low scores (i.e. scores <50) (Figure 6.11).

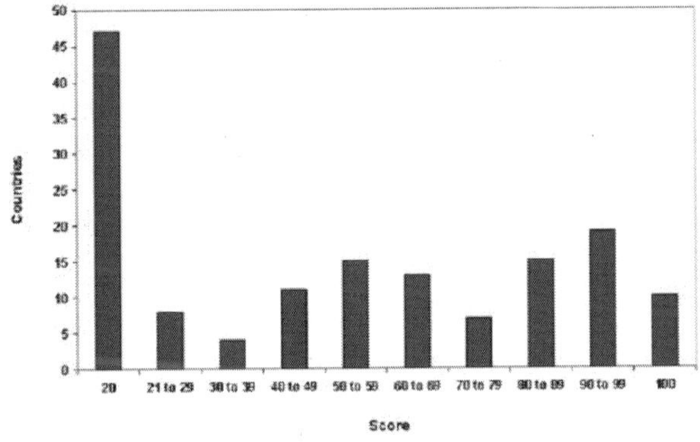

Figure 6.11: Frequency distribution of organism identification scores among 149 inland fisheries countries in 2000

In contrast, there were ten countries that achieved the highest possible score, and an additional 24 countries with relatively high organism identification scores (i.e. score 79 to 99). Finally, the remainder, 35 countries, did moderately well overall at organism identification (scores 50 to 79).

Kenya is an example of a country that scored relatively highly in 2000 with a score of 96. Kenya reported 15 separately identified groups of organisms of which five were identified to the species level. Those identified to the species level accounted for 85 percent of the total production and those identified to Level 4 (genus) accounted for an additional 13 percent of the total production. Thus, the high score was due to identifying nearly all of the production to species and genus levels.

TRENDS IN ORGANISM IDENTIFICATION

Given the importance of organism identification, an important question is the extent to which reporting in this regard is improving. Trends in organism identification were indicated by a calculation of the slope of linear regression line through the scores for the time period. The slope gives a measure of the rate of change in the score and thus provides an indication of the trend in species identification. If scores were available at five year intervals for 1980 to 2000, these were the years used to calculate the regression line making a maximum of five data points in all. For some countries where data for the earlier years were not available, the regression line was calculated for the scores between 1985 and 2000. In all, there were 136 countries for which the trends analysis could be carried out. The resulting slopes were cast into five categories broadly indicative of the trends: marked improvement (21 percent), moderate improvement (23 percent), no change (36 percent), moderate deterioration (13 percent), and marked deterioration (7 percent) (Figure 6.12).

Among those exhibiting no change were 37 countries (27 percent) with scores of 20 indicating that they have continued to report inland production at the least detailed level without any improvement, most of them for the entire period. As an example, Kenya, with scores of 68, 93, 89, 98 and 96 for 1980, 1985, 1990, 1995 and 2000, respectively, is a country that falls into the lower portion of the "Marked Improvement" category. That 59 (44 percent) of the countries exhibit a trend for improvement in organism identification is quite positive, but, overall, there is still much room for improvement in organism identification especially as regards those 37 countries (27 percent) that continue to identify the catch at the least detailed level.

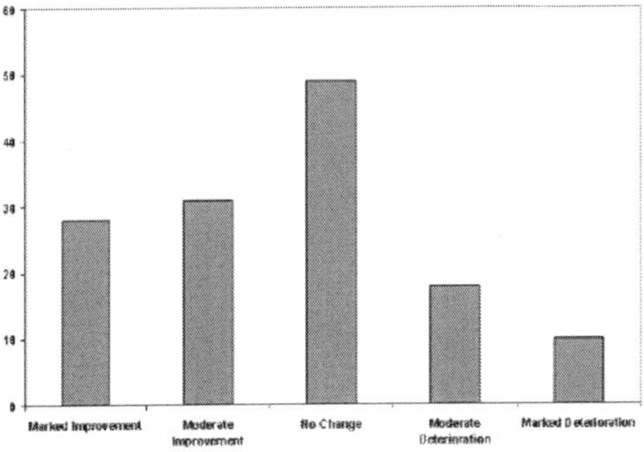

Figure 6.12: Frequency distribution of trends in organism identification among 136 inland fisheries countries 1980 to 2000

THE MOST IMPORTANT SPECIES AND GROUPS IN INLAND FISHERIES PRODUCTION

The most important species and groups in inland production have been aggregated to provide a global overview (Figure 6.13), and views by continents. Globally, only two species, the Nile perch and the Nile tilapia, can be distinguished among the top ten species and groups in production. This is because the remainder is "not elsewhere indicated" (NEI) indicating that many countries have not reported their capture to the species level. Among the NEI groups, tilapias, cyprinids, freshwater siluroids, characins and snakeheads are important. Invertebrates, unidentified mollusks and crustaceans, figure importantly in the world inland catch.

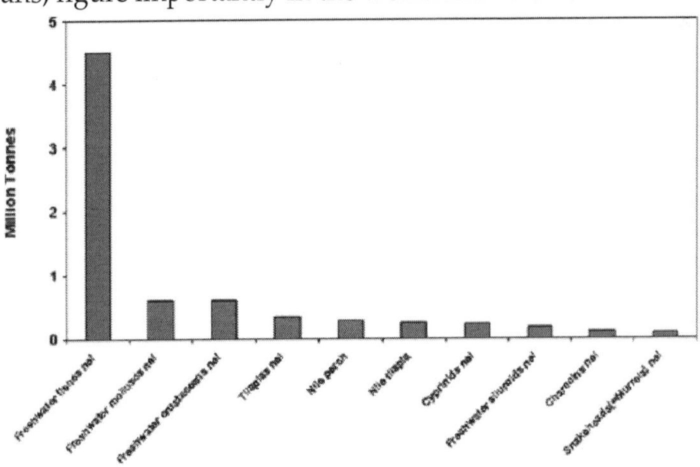

Figure 6.13: All continents - most important species and groups 2001

At the species level Hilsha shad and Kelee shad are among the top identified species in Asia. The tilapias as a group, introduced widely, also figure importantly (Figure 6.14). In Africa, Nile perch, Nile tilapia, silver cyprinid, mudfish, and North African catfish are important at the species level (Figure 6.14). In South America, two species, the netted prochilod and the cachama, are identified (Figure 6.15). In North America, common carp, walleye, and pond smelt are among the important species in temperate and colder regions while the blue tilapia is important in the warmer waters (Figure 6.16). Europe has seven species identified as important, the leading ones being the Azov Sea sprat and the freshwater bream (Figure 6.17). In Oceania, three species are identified among the top ten of which is the Mozambique tilapia, an introduced species, the barramundi, and the short-finned eel (Figure 6.18).

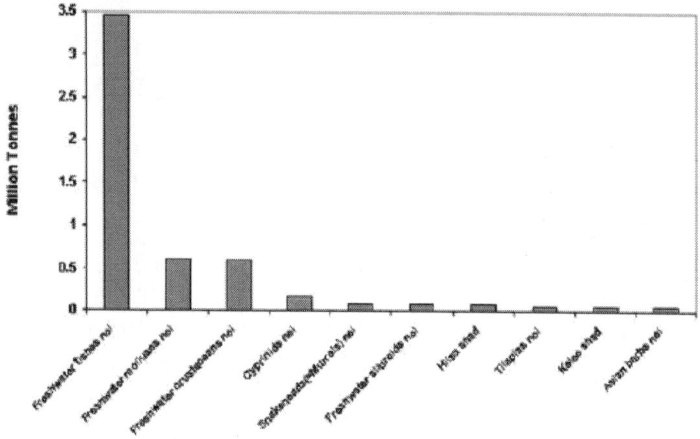

Figure 6.13: Asia - most important species and groups 2001

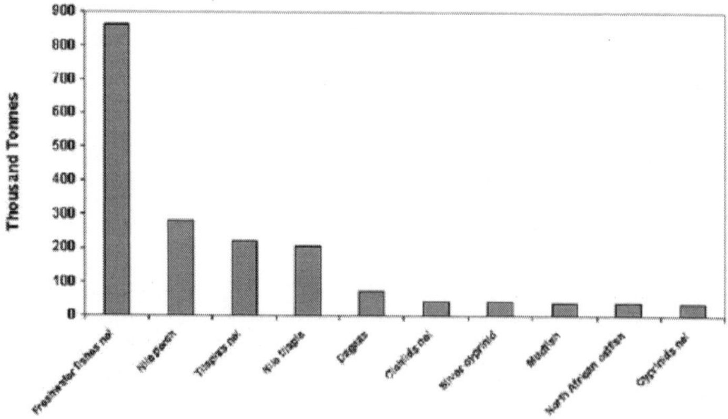

Figure 6.14: Africa - most important species and groups 2001

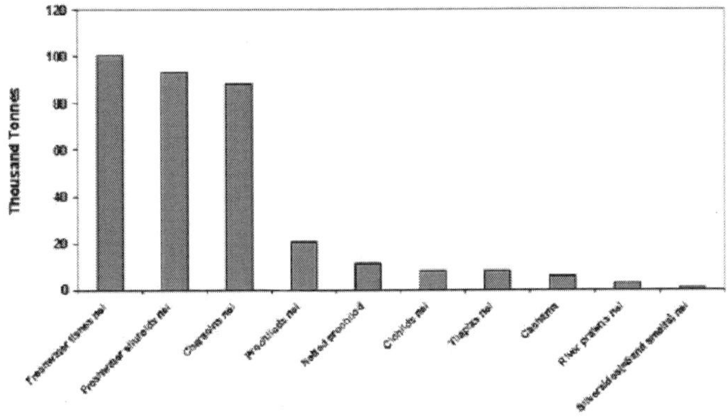

Figure 6.15: South America - most important species and groups 2001

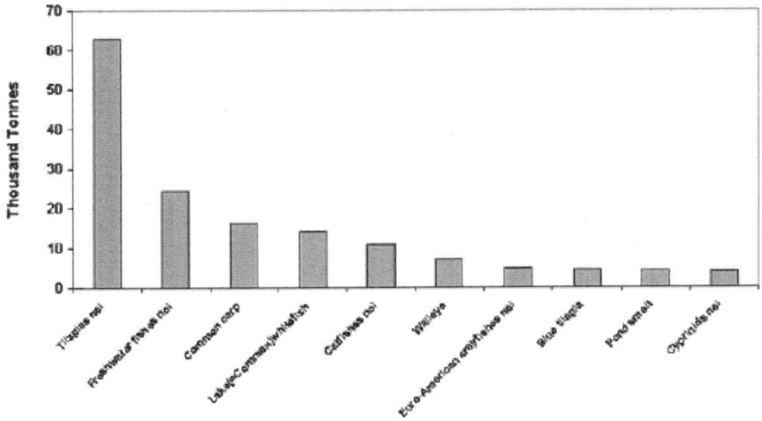

Figure 6.16: North America - most important species and groups 2001

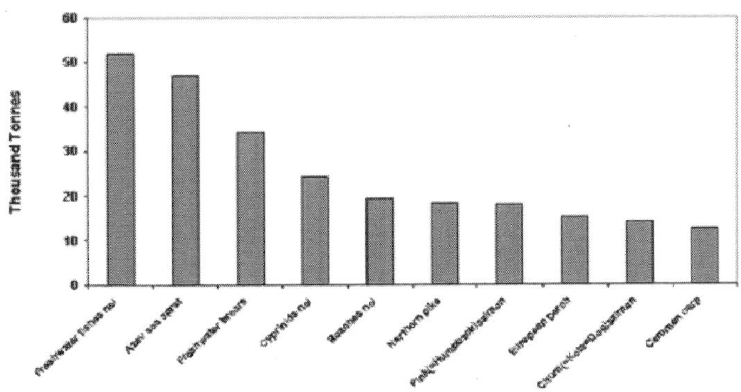

Figure6.17: Europe - most important species and groups 2001

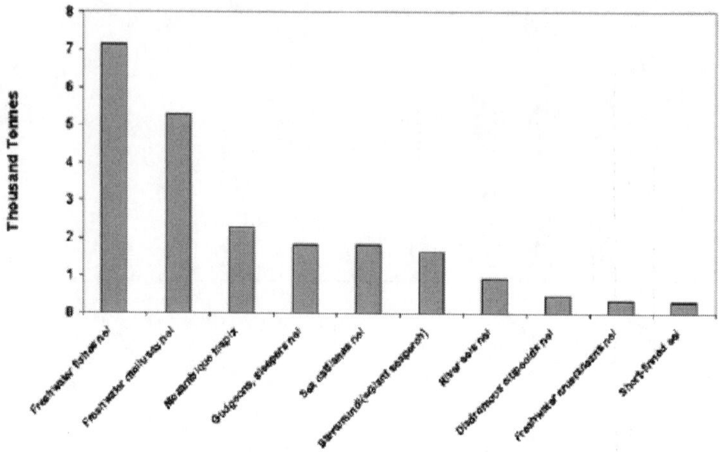

Figure 6.18: Oceania - most important species and groups 2001

7

MANAGEMENT OF INLAND FISHERIES

The world's total production from inland fisheries has increased steadily in the past twenty years. It reached 19 173 642 tons in 1994, comprising just over 20 percent of the world's total landings of all fisheries. Furthermore, inland fisheries are directly related to food security, particularly in Asia and Africa where they provide food, income and employment on a large scale. In 1994, the total fish supply from inland resources in Asia reached 15 279 042 tons, while in Africa it amounted to 1 880 632 tons.

Aquaculture represents some 70% of this production and as such plays an important role in the use of inland resources of Asia and Europe. The overall contribution of aquaculture to fish production is significant and rising whereas the capture fisheries sector is more static.

The problems associated with the management of inland fisheries are very diverse and complex. fish stocks are affected by a number of factors including pollution (eutrophication, toxic wastes, acid rain), river engineering (impoundments, land drainage, flood alleviation), habitat loss through land reclamation and removal of riparian vegetation, afforestation, introduction of non-native species, and overfishing. Thus the management of inland fisheries is different in nature to that of marine fisheries and is basically focused on protection, mitigation and rehabilitation. These are important components of any inland fisheries management strategy. Excluding the large international lakes and rivers for which international cooperation is vital for the management of their fisheries, most management measures relating to thousands of lakes, reservoirs, ponds, rivers and estuaries throughout the world need national initiatives and actions. the management of inland fisheries is invariably part of an overall management strategy of these waters. Multiple use of these waters mean that management objectives have to be prioritized among users.

For example, a dam may provide electricity and water for agricultural purposes affecting at the same time the floodplain and the migratory route of some species. Management strategies are by nature a set of compromised objectives. These objectives are often set by authorities at local, regional or national levels.

International inland fisheries bodies are not expected to manage the inland fisheries of member States. They play the role of promoting a common viewpoint as well as coordinating and facilitating the works of individuals and national institutions on many aspects relating to the management of the inland fisheries. Experiences and views exchanged, shared and discussed are often beneficial and sometimes fundamental to member States in their approach to the management of their inland fisheries.

ROLE OF FAO INLAND FISHERY BODIES

1. European Inland Fisheries Advisory Commission (EIFAC)

i) The present structure of EIFAC

EIFAC was established by the Council in 1957 under Article VI-I of the FAO Constitution (Resolution 2/26). Its area of operation is Europe (FAO Statistical Area 05) and its membership is open to all European Member Nations of the Organization whose territories are situated wholly or partly in the European region. The Commission and its subsidiary bodies are financed and administered by the Organization. The terms of reference of the Commission are:

(1) to assist in the collection and dissemination of pertinent information;

(2) to propose and assist in the organization of appropriate symposia;

(3) to promote liaison and cooperation among governmental organizations;

(4) to advise on the evolution of an organized approach among interested governments of this region toward the development of inland fisheries as may seem desirable and feasible; and

(5) to advise on any other matters appropriate to the promotion of the development and utilization of the inland fisheries within the competence of the Organization.

At its Nineteenth Session held in Dublin, Ireland, in June 1996, the Commission established a new Sub-Commission IV to deal with social and economic issues. Thus the four Sub-Commissions would regroup existing and possible new activities. The Commission, as restructured, has the following subsidiary bodies:

Sub-Commission I:

- Working Party on Fishing Gear Selectivity
- Working Party on Crayfish
- Working Party on Eels
- Working Party on Electric Fishing
- Working Party on Introductions and Stocking

- Working Party on Brackishwater Fisheries and Aquaculture

Sub-Commission II:

- Working Party on Fish Diseases and their Control
- Working Party on Water in Aquaculture
- Working Party on Fish and Crustacean Nutrition
- Working Party on Prevention and Control of Bird Predation
- Working Party on Aquatic Resources Management in Aquaculture

Sub-Commission III:

- Working Party on Influence of Management Practices on the Environment
- Working Party on Effects of Physical Modifications of the Aquatic Habitat on Fish Populations
- Working Party on Accumulated Toxicants in Fish

Sub-Commission IV:

- Working Party on Recreational Fisheries
- Working Party on Legislation Concerning the Aquatic Environment in EIFAC Countries
- Working Party on Education in Fisheries Management
- Working Party on Resolution of Conflicts within River Catchments
- Working Party on Social and Economic Data for Inland Fisheries and Aquaculture

2. Commission for Inland Fisheries of Latin America (COPESCAL)

i) The Present Structure of COPESCAL : COPESCAL was established by the Council in 1976 under Article VI-I of the FAO Constitution (Resolution 4/70). Its area of operation is Latin America and its membership is open to all Member Nations and Associate Members of the Organization which are serviced by the Regional Office for Latin America and the Caribbean. The terms of reference of the Commission are:

(a) to promote, coordinate and assist national and regional fishery and limnological surveys and programmes of research and development leading to the rational utilization of inland fishery resources;

(b) to assist Member Governments in establishing the scientific basis for regulatory and other measures for the conservation and improvement of inland fishery resources; to formulate such measures through subsidiary bodies as required, and to make appropriate recommendations for the adoption and implementation of these measures;

(c) to encourage efforts on a national basis and, when appropriate, to coordinate such efforts on a regional basis with a view to preserve the quality of water;

(d) to assist in the development of aquaculture and stock improvement, including the control of fish diseases, the importation and introduction of exotic species;

(e) to assist in appraising the economic and recreational values of inland fisheries and their development;

(f) to promote and assist in the utilization of the most effective fishing craft, gear and techniques;

(g) to promote and assist with activities for the processing, preservation and marketing of fish and fish products;

(h) to encourage education and training through the establishment or improvement of national and regional institutions and by the promotion and the organization of symposia, seminars, study tours and training centres;

(i) to assist in the collection, interchange, dissemination and analysis of data related to inland fisheries;

(j) to assist Member Governments in formulating national and regional programmes to be implemented through sources of international aid to help achieve the objectives referred to in the preceding paragraphs;

(k) to perform such other functions as are related to the rational management and development of inland fisheries in the Region.

The Commission has the following subsidiary bodies:

- Executive Committee
- Working Party on Fishery Resources
- Working Party on Fishery Technology
- Working Party on Aquaculture

3. Committee for Inland Fisheries of Africa (CIFA)

i) The Present Structure of CIFA : CIFA was established by the Director-General in 1971 under Article VI-2 of the FAO Constitution following the decision of the Council at its Fifty-Sixth Session (Resolution 1/56). Its area of operation is Africa (FAO Statistical Area 01) and its membership is selected by the Director-General from African Member Nations and Associate Members of the Organization on the basis of their active interest in inland fishery development in Africa and of their potential contribution to the effective discharge of the functions of the Committee. The terms of reference of the Committee are:

(a) to promote, coordinate and assist national and regional fishery and limnological surveys and programmes of research and development leading to the rational utilization of inland fishery resources;

(b) to assist Member Governments in establishing the scientific basis for regulatory and other measure for the conservation and improvement of inland fishery resources, to formulate such measures through subsidiary bodies as required, and to make appropriate recommendations for the adoption and implementation of these measures;

(c) to promote and coordinate efforts on a national and regional basis to prevent damage to the aquatic environment, including the prevention and control of water pollution;

(d) to assist in the development of fish culture and stock improvement, including the control of fish diseases and the importation of exotic species;

(e) to promote and assist in the utilization of the most effective fishing craft, gear and techniques;

(f) to promote and assist activities concerned with the processing, preservation and marketing of fish and fish products;

(g) to encourage education and training through the establishment or improvement of national and regional institutions and by the promotion and the organization of symposia, seminars, study tours and training centres;

(h) to assist in the collection, interchange, dissemination and analysis of statistical, biological and environmental data and other inland fishery information; and

(i) to assist Member Governments in formulating national and regional programmes to be implemented through sources of international aid to help achieve the objectives referred to in the preceding paragraphs.

The Committee has the following subsidiary bodies:

- Sub-Committee for the Development and Management of the Fisheries of Lake Victoria
- Sub-Committee for Lake Tanganyika
- Sub-committee for the Protection and Development of the Fisheries in the Sahelian Zone
- Working Party on Aquaculture
- Working Party on Stock Assessment
- Working Party on Floodplain Fisheries
- Working Party on Pollution and Fisheries

4. APFIC Aquaculture and Inland Fisheries Committee (AIFIC)

AIFIC was established as a result of restructuring of APFIC in 1993 and has not held any session yet. It is a subsidiary body of APFIC operating under Article XIV of the FAO Constitution. The terms of reference of AIFIC are:

a) to keep under review the state of these resources, of the industries based on them and of their environment;

b) to formulate and recommend measures an to initiate and carry out programmes or projects to:

   i) increase the efficiency and sustainable productivity of inland fisheries and aquaculture;

   ii) conserve and manage resources;

   iii) protect resources from pollution;

c) to keep under review the economic and social aspects of fishing communities and recommend measures aimed at improving the living and working conditions of those involved in inland fisheries and aquaculture and otherwise at improving their contribution to social and economic goals;

d) to encourage, recommend, coordinate and undertake, as appropriate, research, training and extension activities in all aspects of sustainable inland fisheries and aquaculture;

e) to assemble, publish and disseminate information regarding the living inland aquatic resources, fisheries based on these resources and aquaculture;

f) to direct and provide guidance to the relevant Working Parties established under this Committee; and

g) to carry out such other activities as may be necessary for the Commission to achieve its purpose as defined above.

The Committee has the following subsidiary bodies:

- Working Party on Aquaculture and Inland Fisheries
- Joint Working Party on Fishery Statistics and Economics
- Joint Working Party on Fish Technology and Marketing

The two Joint Working Party will also serve APFIC Committee on Marine Fisheries.

(iii) Strengthening of FAO Inland Fishery Bodies

Strengthening of FAO inland fishery bodies does not require fundamental changes to the present structure and functions of these bodies. They need an increased financial and resources support by their members in addition to those already provided by the Organization.

DATA REQUIREMENTS FOR INLAND FISHERIES MANAGEMENT

Any resource can only be managed if sufficient good data and information are available. Unfortunately this principle has been generally overlooked in the case of inland fisheries where data are essentially weak and generally insufficient for taking the kinds of decisions that are needed. This lack has contributed to the generally poor state of inland fisheries resources around the world.

There is, therefore, a general need to collect more and better data and information on the inland water environment, the fish and the fishers that form the essential components of a fishery. Collection of data and information is expensive in that it requires people, transport and communication systems to be effective.

Furthermore it requires trained people that may not always be available. Because of the cost and the demands on often scarce trained personnel it is important that any programme for the collection of fishery information be efficient. It is also important not to collect more data than is needed for the management of the fishery. Especially in financially constrained circumstances the data and information collected should be tailored to the job on hand.

This requires a clear definition of the objectives of management and knowledge of the functions and limitations of the different types of information. This chapter identifies some of the types of information and data required for management at different levels of complexity and for different purposes.

Purposes of data and information collection

Information and data about a fishery are collected to support the objectives and programmes of whoever is managing the fishery. There are a number of objectives for management, which are not always shared by all stakeholders in the fishery, whose priorities may often differ according to their interest. Objectives of inland fishery management include:

Extractive objectives, such as fish protein supply or recreational opportunities;

Social objectives, such as income, equity of distribution of benefits, reduction of social conflict;

Fiscal objectives, such as revenue and foreign earnings; and

Conservation objectives, such as sustainability or biodiversity conservation.

Most fisheries are managed for a mix of these objectives. It is, therefore, important to agree on a common approach to the fishery so that data collection systems serve the greatest number of interested parties.

Types of data and information

A distinction is made in this paper between data and information. Data are raw numerical values (statistics) that must be analysed to become information. Information may also be non-numerical and deal with such questions as whether it is the men or women who fish or whether a particular species migrates. In both cases the interpretation process is as important as the collection process. Data and information for various sources must be put together to form the basis for management.

Data can be:

1) Absolute, which attempts to assign values to such variables as catch that are as close to reality as possible. Absolute values are generally a snapshot of the situation at a point in time.
2) Relative, which attempt to trace the movement of values over time from a baseline. The baseline itself may be an absolute value or one that is selected arbitrarily in the time series.

Subjects of information

Information can be obtained about: The fish, including data and information on feeding, breeding, growth and mortality as well as any migrations and other aspects of biology; The fishery, including statistics on landings and effort, information on gear and its performance, information on behaviour and income of the fishers and their families; The environment, including information on water quality and quantity and the degree of modification of aquatic habitats and ecosystems.

Timing and location of data collection

Information and data can be obtained from:

1) One-off studies and surveys that are expensive and detailed operations usually designed to establish absolute magnitude of a resource for valuation of a fishery, to obtain essential livelihoods information on the social and economic situation of fishers, to establish baseline information and to set up a sampling frame. A frame survey divides a river or lake into various strata or zones according to ecological, fishery or social characteristics, identifies sampling locations that are typical of the various strata, and lays down a schedule for the timing of sampling. Division of the area into strata makes it easier to group sampling sites and interpret the data collected from them.
2) Regular surveys are carried out at regular intervals to monitor changes following major studies. A sampling frame may be used to select the number and location of the sampling sites.

   Daily/weekly collection of data is used to establish local trends at individual markets and landings. A sampling frame may be used to specify landings and markets that are main data collection points.

Research programmes that address particular issues or aspects of the fishery and its environment. Research is needed, for example, to define details of the biology and ecology of the living organisms involved in the fishery, the social and economic circumstances of the fishers and their families, market patterns and conservation needs.

3) Existing data and information should never be overlooked. With the exception of the very largest lakes and rivers, inland fisheries are replicated systems based on many similar small lakes, streams and reservoirs. As a result, there is an accumulation of data and information internationally that has been codified in series of textbooks, guidelines and indexes that can be transferred to a national fishery with relatively little effort. Within a country there is also much information on aspects of the fisheries in a particular basin that can contribute to establishing a historical baseline and measure of change.

Data can be collected from:

1) Markets, which are the great concentration point of the fish sold and passing through the commercial network. They are particularly useful in diffuse fisheries where there are no defined landings. Because of the volume of fish passing through, and the fact that the product has often been treated by smoking, salting or drying, they can be relied upon only for coarse statistics on weight of the fish and maybe main species types. Market surveys carry the danger of double counting, as the same fish may appear more than once in the same market or in successive markets in the chain.
2) Landings, which are the primary collection point of fish landed. These are particularly valuable in lakes or large rivers where definite points have developed for landing and sale of fish. They collect data primarily on quantity of fish landed by the fishery and can also be used to gather data on species, weight, length and sexual condition. They are also good for the collection of data on the types of gear in use. Landing statistics are less applicable in some river fisheries where landing and sale are diffuse and the majority of the fish caught is not channelled directly to discrete localities. Landing data are also incomplete in that it misses the often considerable amount of fish consumed by the fishers, their families and associated communities (autoconsumption).
3) Fishing villages, through family and household surveys, which can detect sources of fish that are not dependent on landings and boats but which are collected from small water bodies by women and children. Such surveys are also useful for understanding how the fishery works. They can supply data on gear use patterns and on the social and economic organization of the fishers. Individual fishers, who are difficult to sample on a regular basis in

most fisheries. They are however the best source of data on the performance of individual gear use and on the types and sizes of the fish caught. They can also be helpful in providing information about autoconsumption patterns and the allocation of work among participants in the fishery.

COMPLEXITY OF INFORMATION:

Data can be divided into:

1) Indicators, which are criteria, species or processes that are identified as representing changes to the fishery. These can be values, changes in quality or quality, trends in relative values or a simple presence or absence of a condition or species. Indicators are usually related to reference points or values that signal critical changes in the fishery. Indicators should be chosen so they are relatively easy to sample and calculate from routinely collected data. Reference points are more difficult as they depend on the management model to be adopted or the policy regarding the fishery as a whole.
2) Variables, which are the components of indicators and are the actual material collected. Which variables are collected are determined by the particular management model used and by logistical considerations such as ease and cost of collection and the availability of trained people.
3) Indexes, which are combinations of variables into simple models that are based on information collected over a period of time and a wide geographic area. They serve primarily as a basis for comparison of the situation in a particular fishery with a wider standard. For example indexes such as the morpho-edaphic index or the river catch index can be used to determine approximate values of the catch to be expected from lakes and rivers and integrated biotic indexes (IBI) can be used to monitor the health of the fish stock.

Detail and exactness of data

Accurate data is that which closely corresponds with the true situation of the fishery. However, data can be collected at differing levels of detail. For example catches can be calculated precisely to the second or third decimal point but such a degree of precision means very little because the fishery is constantly changing and such a level of detail has little influence on the decisions to be made. It is generally more important to be accurate and arrive at a more general figure of catch that actually influences policy.

The type, amount and degree of detail of the data collected depends on the type of model used for management of the fishery and governmental policy on the degree of intervention in the management of the fishery. In resource poor situations it is better to adopt simpler models that require less data. It is better to collect less data that is accurate than to expend effort collecting large amounts of poor quality information in the hope that the extra detail adds more information. In general, however, the more complex the decision to be made the more information required to make the decision in an informed manner.

For example, a decision regarding whether or not to establish a sampling site at a particular landing can be made simply on the basis of a general knowledge of catch in the area that can even be obtained from a simple index. Decisions regarding the number of individual catch quotas and licenses at the same landing could require detailed knowledge of the performance of the fishery and the social and economic circumstances of the fishers. Whatever the circumstance there should be a long-term commitment to collect at least the most basic catch statistics at representative landings.

Models of the fishery

The following two models of fishery management illustrate the differences in data needed by different models.

Maximum sustainable yield (MSY)

The maximum sustainable yield concept provides a reference point that predicts the level of effort at which the maximum amount of fish can be sustainably captured from a stock (Fig. 1). MSY tends to be an absolute value. There are two approaches to this model. The first is a simple approach based on an analysis of catch against effort and requires data on both these variables. This is, therefore, a relatively easy model to use and requires very little data. However, it can only be done on a single species basis and so the same two variables have to be collected for all species in the fishery.

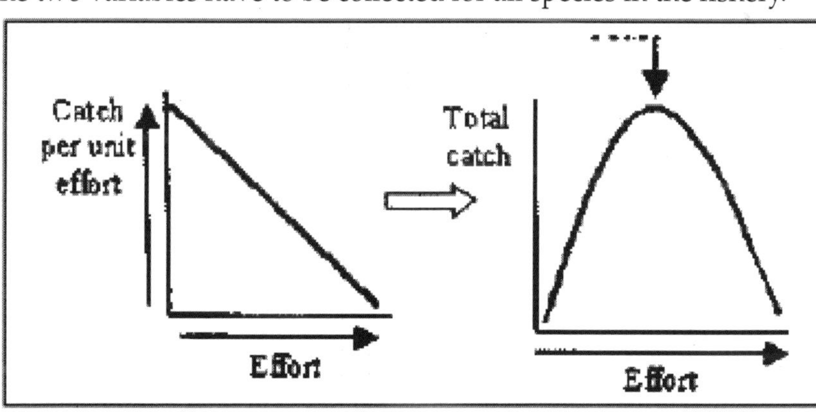

Figure 1 Diagram of MSY model

This means that it is a useful model for lakes where few species make up the majority of the catch. It is equally useful in river fisheries that concentrate on a few large species. It is less useful in true multi-species fisheries. The second method provides better estimates of MSY but can only be obtained through more complicated stock assessment procedures that require complete data on mortality, recruitment and growth for each species.

Such complicated models allow for more detailed evaluation of management regimes, such as those based on size limits, but do not necessarily provide better estimates of MSY.

These types of models are also known as surplus production models (SMP) and are only of use when there is plenty of contrast in the data. This means that the fishery is often overexploited by the time the trends to maximum sustainable yield become clear.

Information and data about a fishery are collected to support the objectives and programmes of whoever is managing the fishery.

Multi-species fisheries

The multi-species (Fig. 2) model predicts the behaviour of complex assemblies of fish when subject to fishing and environmental pressure.

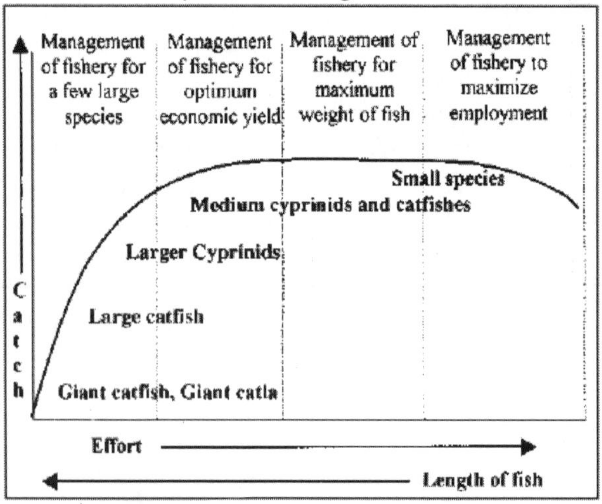

Figure 2. Diagram of multi-species model

It shows a total yield that is sustained for a considerable range of effort giving a flat or plateau type yield curve. This sustained yield hides a series of changes to the fish community whereby successively smaller species replace the larger species.

The model tends to be relative as it does not predict the level of sustained yield but does trace the health of the fish assembly as a series of measures (Table 1).

Table 1 Criteria used in assessing the state of health of multi-species fisheries. Trends indicate a fall in the general health of the fish assemblage.

Indicator	Trend
Level of catch	• Falling levels of total catch levels in single species fisheries, • Catch levels can be maintained over a wide range of effort
Mean length	• Disappearance of larger fish and falling mean size within species • Disappearance of larger species and falling mean size of catch as a whole
Number of species	• Initially rise in number of species in catch from a few large to many small ones • Later falling numbers until the fishery is confined to few very small species
Type of species	• Decline and disappearance of anadromous and long distance riverine migrants • Decline and disappearance of native species • Rise in numbers of exotics where introductions have occurred * Decline and disappearance of higher trophic levels (predators) • Decline and disappearance of species with high oxygen requirements
Response time	• Shortened time between flood events and response by population
Other indicators:	• P/B ratios rise • Mortality rates(Z andf) increase • Higher incidence of diseased and deformed individuals

Many of these measures are relatively simple and easy to obtain such as the mean length of the fish. Assessment of the state of complex fish stocks can probably be done on the basis of mean length alone. Other measures are more difficult to obtain such as the Production/biomass ratio (P/B) that measures the total biomass present and the rate of production by the fish community or species.

One problem with this model is that both fishing and environmental factors influence the fishery in a similar manner. It is, therefore, important to be able to disentangle the relative contribution of the fishery and the environment. This generally requires better knowledge of environmental states than is often available.

INFORMATION FOR SPECIFIC PURPOSES

The following section examines examples of the decisions that have to be made regarding the fishery and the type of data required.

National decision-making

National decision making on the fishery usually concerns an assessment of the value of the sector to the national economy, the allocation of funds to the sector or the amount of funds that can be obtained from it in the form of taxes and license fees.

Management of fish stocks

The classical approach to management of fish stocks is by centralized control of the fishery through mesh size restrictions, prohibitions on gear, closed seasons, harvest reserves and restrictions of access through licenses or catch quotas. Classical fisheries management requires very complete data on the fish, the effort and the selectivity and performance of the fishing gear used. Completeness of data of this order is needed for all major species in the fishery. All too often the data needed for such decisions was lacking and the national authorities in charge of the fishery assigned arbitrary values to such factors as mesh size.

This led to a general disregard for regulations on the part of fishers and is a contributing factor to the poor state of inland fisheries today. More recent approaches tend to concentrate mainly on restriction of effort and co-management involving the participation of other stakeholders in the fishery in decision-making and management. The information needed for this approach is far simpler and qualitative, length-based data often suffice. Furthermore the audience for the information is different. In centrally controlled systems only the government planning authorities need the information. In more recent systems the information also needs to reach the stakeholders to assist them in local decision-making.

In addition to the output and input controls traditionally applied to the fishery, fish stocks are increasingly manipulated by such processes as stocking or introduction of new species.

Fishers and their communities

Fisheries management has moved away from an early preoccupation with the fish to a greater concern about the livelihoods of the fishers and their families. This increased social and economic dimension has given rise to the need for other categories of information than the simple fisheries statistics that formed the traditional basis for management. There is increasing concern with such issues as equitable distribution of the benefits from the fishery, reduction of conflict both within fishing communities and between the communities and other users and gender participation in the fishery.

Conservation and rehabilitation

There is increasing concern about the conservation of inland fish resources. In part, this is a response to the contractual obligations of the individual countries to the Convention on Biological Diversity, and any other conventions such as Ramsar to which they might be signatory. It is also a response to the perception that fish stocks are in decline everywhere and large species particularly are disappearing and may well become extinct.

CONCLUSION

This chapter suggests some principles for planning, collecting and interpreting data and information.

First, data collection should only be pursued with a clear view as to what the data is to be used for. Fishery management policies and objectives should be defined and the models for management chosen. Reference points and variables need to be chosen to guide the policies and support the models. The data and information needed for the variables can then be selected.

Second, in resource poor societies the data collected should be limited to that which can be collected with the resources available while conserving the quality of the data. Accuracy of data should be favoured over precision. Policies and models should not be overly complex and should be consistent with the data gathering capacity of the society. Third, data on their own are of little use and must be analysed to become information. The process of analysis is as important as that of data collection.

Fourth, whatever the circumstance there should be a long-term commitment to collect at least the most basic catch statistics at representative landings. Many institutions in a modern state generate fisheries related information. Central or regional government agencies are usually charged with the collection and interpretation of basic fisheries data. Research on particular topics is usually the function of specialized research institutes and programmes in government or in universities. Externally funded projects may carry out particularly intensive one-off studies such as frame surveys. To this can be added the wealth of traditional knowledge held by the fishers and other stakeholders in the fishery. There is also usually a considerable amount of information already available in the world's libraries that can contribute to our knowledge. The problem you are facing is probably not new and has already been studied by others, so the answer may lie somewhere in the literature.

Information is only of value when it is used. The information from all sources has to be interpreted and fed into the management plan. One problem that the fisheries sector has faced is the poor quality of its data as compared to other users of aquatic resources. A further problem lies in the exclusion of certain types of information gathering mechanisms from the management process such as universities or traditional knowledge of fishers.

This means not only that fisheries management has been generally defective, but also that the sector has not been able to secure an adequate share of the water on which it depends. This situation needs to be rectified by the improved development of sampling and other information gathering activities as well as the development of mechanisms that bring together the information from the various national and international sectors.

UN urges management of inland fisheries for food, income stability

The need to conserve and properly manage inland fisheries is often neglected despite the crucial role they play in balancing aquatic ecosystems and as sources of food and livelihoods to millions of people across the world, a new report released today by the United Nations Environment Programme (UNEP) says.

According to the new publication, rapid environmental changes constitute a challenge to the viability of future fish stocks despite more than 40 years of steady production globally.

Changes in seasonal flooding patterns and loss of habitat, unsustainable farming practices and excessive use of water from lakes and rivers are some of the threats to inland fish, according to the report, prepared jointly by UNEP and the World Fish Centre and entitled Blue Harvest: Inland Fisheries as an Ecosystem Service.

Urbanization, road-building, pollution, including wastewater discharge into inland fisheries and climate change are some of the other threats to the fresh water fish, says the report, which was launched at the 10th conference of the parties to the UN Convention on Biological Diversity (CBD) under way in the Japanese city of Nagoya.

It stresses that aside from the nutritional benefits, fish also play a key role in the functioning of aquatic ecosystems and cites the consumption of plankton, plants, insects, and other fish as critical to the stability and resilience of river and lake habitats.

Nutrients and organic matter from fish eggs, carcasses and excretion support the production of algae, insect larvae and other fish species in rivers and lakes, helping to link ecosystems.

Declining fish populations can have a serious knock-on effect for other organisms, the report says, giving the example of the widespread deaths of the cisco fish in Lake Mendota in the United States which led to changes in the plankton composition of the lake, decreased the level of nutrients in the water column and caused a decline in the biomass of algae. "This fascinating report has brought to the fore the often neglected subject of inland fisheries. While marine fisheries are under increasing scrutiny, those based on river and lake systems rarely engage the international community - an oversight of potentially profound implications," said Achim Steiner, UNEP's Executive Director.

"Why? Because an estimated 100 million people in Africa alone get important levels of daily protein from these inland sources alongside essential vitamins and minerals. Meanwhile, unofficial estimates put the global inland catch at close to 30 million tonnes, comparable to official marine catches, and employment at 60 million people - 13 million more than in equivalent marine fisheries," he added.

The report urges States to adopt an 'ecosystem approach' to managing inland fisheries because of the multiple threats to their health and productivity.

Such an approach needs to take into account a wide range of factors, including reducing pollution and destructive fishing practices, maintaining river flows and restoring damaged habitats and protecting wetlands and other feeding and fish spawning grounds.

It recommends that new dams be located where they have the least impact on river ecosystems and that fish-friendly designs that are managed to allow fish migration and delivery of seasonal flows. Older dams need to be altered to provide similar benefits, the report adds.

"Recent achievements in the United States and the Vu Gia-Thu Ban River basin in Viet Nam show that political will and careful planning can provide win-win solutions," said Patrick Dugan, the lead author of the report, who is based at the World Fish Center in Penang, Malaysia. "These have kept some river corridors free from dams, while others are managed for both environmental and hydropower objectives. We need urgently to replicate these successes more widely and in larger rivers if we are to sustain the world's inland fisheries."

8

ANTIBIOTICS IN SALMONID AQUACULTURE

As in other agricultural sectors, antibiotics can be used in fish to effectively treat bacterial diseases. However, aquaculture in BC is unique amongst the food-producing industries as antibiotics are administered by veterinary prescription. Antibiotics used in aquaculture are administered to fish in feed to treat disease - not to promote growth.

Like land-based forms of raising livestock, where large numbers of animals are placed in a very limited space, intensive fish farming provides various diseases and parasites ideal conditions to spread. Infectious diseases like furunculosis were previously treated with large quantities of antibiotics added to feed. In our discussion of the environmental impacts of antibacterial agents in fish farming, we would remind the reader that the use of antibiotics in the Norwegian fish farming industry has fallen from 50,000 kilograms to 500-600 kilograms in fifteen years (Grave et al., 2002).). According the Norwegian Institute of Public Health statistics the antibiotic use in fish farming was at its lowest in 1999 with 591 kg. Since that time the use has increased to a top in of 1,428 kg in 2006 and a reduction to 649 kg in 2007.

Current levels are low enough not to pose any environmental problem. Nevertheless, there are two reasons why we wish to focus on this issue. First, the use of antibiotics is still high in other countries where salmonids are farmed. This applies to Chile, for example, a country not as advanced in preventive fish health as Norway. There is no vaccine against the disease piscirickettsiosis, the disease causing the greatest losses in Chilean fish farming (Olsen, 1999). Antibiotic use in Chile is estimated at fifty tonnes (Jensen, 2002; Nutreco 2002). Second, we fear that consumption in Norway will rise if the farming of other species such as cod reaches significant production volumes. The reason for the rapid decline in Norwegian antibiotic use is primarily due to the availability of effective vaccines against the chief bacterial infections, as well a generally better fish health as a result of more suitable operation locations, less stress and improved hygiene against contagions (Poppe, 1999).

New farmed species will pose new health challenges, and the development of new vaccines may take time (Bleie, 2002). New salmon diseases and new variants of known diseases may also help to reverse the favourable trend in Norway. In 2002 consumption increased to about 1,000 kilograms, especially due to problems with winter ulcers (Norwegian Animal Health Authority, 2003; Rasmussen, 2002).

ENVIRONMENTAL IMPACTS OF ANTIBIOTICS IN FISH FARMING

Antibiotics are administered to fish through medicated feed that is mixed at feed plants. Antibiotics enter the marine environment when some of the feed is not eaten and sinks to the seabed or is eliminated in fish excretion (Smith, 1996). Three effects of the spread of antibiotics to the marine environment are highlighted:

- Antibiotic resistance
- Spread to wild fish
- Retarded decomposition of organic material

Antibiotic Resistance

After intensive, repeated medication using the same antibacterial agent, reduced efficacy is often seen after a time against the disease in question. Before vaccination programs largely eliminated furunculosis outbreaks, oxolinic acid was chiefly used to treat it. After repeated treatments, it was noted that the efficacy was reduced, and in 1999 an investigation showed that 36% of the furunculosis bacteria were resistant against one or more antibiotics (Sørum, 1999). As a consequence of the emergence of resistance, treatments must be repeated or new agents employed. However, antibiotic resistance is not purely a problem of fish health. Resistance can develop in other bacteria in the marine sediments and in the water, thereby spreading to other organisms in the marine environment (Sandaa et al., 1992). The total burden of resistant bacteria people are subjected to is growing and may create new problems for human health.

Spread to Wild Fish

Antibiotics are also spread to wild fish directly when cod, for example, eat medicated feed that falls through the cages. This fish, in turn, may be caught and eaten by people, who thereby ingest limited doses of antibiotics (Røstvik, 1997).This is undesirable, when one considers the development of resistance in people. Since the use of antibacterial agents is currently so small, the quantity that wild fish ingest from feed spills will be practically equal to zero.

RETARDED DECOMPOSITION OF ORGANIC MATERIAL

Antibacterial agents, as the name indicates, are meant to kill bacteria. But as was mentioned above, parts of the doses of medication end up in the marine environment, especially in the sediments beneath and around the fish farm.The consequence of the spread is that the number of bacteria in the sediments is also reduced. Hansen et al. (1992) found that the total number of bacteria in the sediment after the addition of the antibacterial agents oxytetracycline, oxolinic acid and flumequine by 50-70% respectively. A fish farm has considerable discharges of organic material in the form of unutilised fish feed and excrement (see below in this chapter). A sharp decline in the bacterial flora around the farm retards the decomposition of the organic material and contributes to accumulation.

The Biodegradation of Antibacterial Agents in Sediments

The biodegradation of antibiotics in the sediments goes relatively slowly (a half-life of up to 150 days in the topmost sediment layer, 0-1 cm), but varies substantially among the various agents and among types of sediment. A study in which samples were taken up from under various fish farms indicates that Florfenicol has a half-life a fraction of that of the other common agents (Hektoen et al., 1995)

Important to Focus on Antibiotics

The environmental impacts of high levels of antibiotic use in fish farming will be substantial and unwanted. Therefore, it is good news that the use of antibiotics in the Norwegian fish farming industry has nearly been eliminated in fifteen years, from 50 tonnes to about 1,000 kilograms. Nonetheless it is important to keep the focus on this issue in countries where the development of vaccines and general preventive health efforts have not come as far, and where considerable antibiotic use continues to be maintained.

Monitoring and Control

Health Canada (Bureau of Veterinary Drugs) licenses all veterinary drugs through the *Food and Drug Act*. Provincially, the Ministry of Agriculture monitors sale of veterinary drugs under the *Pharmacists, Pharmacy Operation and Drug Scheduling Act* and *Medicated Feeds* regulations. As all antibiotics used in salmon (except broodstock which are not used for food) are administered through feed, monitoring the sale of medicated feeds provides the Ministry with information on antibiotics administered to farmed food fish. The Ministry of Agriculture maintains a database of this information for aquaculture.

Facts about Antibiotic Use in BC Salmon Aquaculture

- There are currently four antibiotics licensed by Health Canada for use in aquaculture.
- On average approximately 2.5% of all milled feeds are medicated each year.
- Most of the antibiotics are used in fish less than 2000 grams in size (non-food fish size).
- There are no hormones or pesticides used on food fish in the BC aquaculture industry.
- Antibiotics used in BC Aquaculture are veterinarian prescribed.

Food Safety

All therapeutants have a withdrawal time - a period of time after treatment ceases after which the fish are considered drug free. Withdrawal times vary depending on the therapeutant used and other factors such as water temperature. Fish cannot be harvested for food before the withdrawal period is over.

The Canadian Food Inspection Agency (CFIA) is responsible for monitoring food safety and audits all agricultural meat products for the presence of drug residues. CFIA conducts spot audits on farmed fish for the presence of drug residues.

Antibiotic Use in Salmon Farming Compared to Other Uses

With few exceptions antibiotics are used routinely in food animal populations in Canada. The possible negative consequences of widespread use of antibiotics in agriculture - especially to promote growth is under question. Over-use and inappropriate use of prescription drugs by the human population is considered by some to be a looming issue. In contrast, therapeutant use in aquaculture is used only to treat disease, judicious use of antibiotics in the aquaculture industry has been the norm and reducing antibiotic use remains a goal of all aquaculture producers.

EXCESSIVE ANTIBIOTICS

Large volumes of antibiotics are used in salmon farming to treat disease. Open net-cage aquaculture systems encourage antibiotic use because farmed fish are fully exposed to diseases and parasites that occur in the ocean environment.

With the combination of exposure to disease and the ability of disease pathogens to multiply quickly in the high density conditions common to net-cages, excessive and preventative antibiotic use is common practice in the salmon farming industry.

Farmed salmon are treated with antibiotics through medicated baths and medicated food. By law, a "withdrawal period" – a set number of days between the last use of the antibiotic and the harvest of fish for human consumption – must be met in order to limit the residues of antibiotics in the final product to safe levels.

The major cause of concern with the use of antibiotics in farmed salmon (and other livestock) is that many of these antibiotics are also used to treat human diseases. Frequent use of antibiotics in aquaculture and other industries poses a risk to human health by allowing disease microbes to become resistant to antibiotic treatments – making it more difficult to treat human disease.

Salmon Aquaculture Chemical Report

The multistakeholder World Wildlife Fund (WWF) Salmon Aquaculture Dialogue commissioned a report on chemical use in salmon farming.[1] The committee of expert scientists that authored the report explain that:

"...this use of large volumes of antibiotics can only be explained by excessive and prophylactic [preventative] use. Excessive and prophylactic use of antibiotics in animal husbandry is in general the result of shortcomings in rearing methods and hygienic conditions that favor animal stress, and opportunistic infections and their dissemination."

The committee of scientists, including a research scientist from Fisheries and Oceans Canada marine environmental sciences division, raised concerns about the large quantities of antibiotics that are applied in Chile and in BC. The quantity of antibiotics prescribed per metric ton of production is significantly higher in comparison to Norway or Scotland.

In open net-cage fish farming it is likely for antibiotics to pass into the environment, affecting wildlife remaining in the environment for extended periods of time. The report concludes that "antibiotic-resistant organisms in the marine environment will, in turn, pass their antibiotic resistance genes to other bacteria, including human and animal pathogens." The whole ecosystem (including fish, shellfish, marine mammals, and human beings) is affected.

The industry continues to rely on these treatments, administered in net-cages open to the ocean, despite growing concerns over antibiotic resistance.[2]

Learn about closed containment systems and how they may offer a promising solution to the excessive use of antibiotics.

Antibiotics and Salmon Farming

Developing new vaccinations will help Tasmanian fish farmers reduce their reliance on antibiotics, according to an expert who's worked in the industry here and overseas for 20 years.

David Whyte is a marine biologist and Huon Aquaculture's production planning manager.

He says antibiotics will always have a role in fish farming as a last resort to keep fish health, but vaccinating fish against well known problems has allowed salmon farmers overseas to significantly reduce their use of antibiotics.

"In Norway and Scotland, the industry's been going longest in those two countries, so most of the fish are covered by vaccines against the types of diseases they have there.

"Here we're quite lucky, in Tasmania, because we have relatively few diseases and they're generally a consequence of the different environment, the higher water temperatures down here.

"The things about having individual issues in other countries is that sometimes you can't leverage directly off the solution that they're applied in another country.

"The problems are different [but] the solutions are common," he says.

And some Tasmanian vaccines have already been developed.

"There are some vaccinations for some issues that were found a few years ago in Macquarie Harbour, so in the west coast sites, there's a couple of conditions up there that vaccines were developed successfully in Tasmania, by Tasmanian scientists.

Problems Associated with Salmon Farming

There are a number of environmental issues associated with salmon aquaculture:

Interactions with wild fish: When fish escape from farms and survive in large numbers or establish their own breeding populations, they will compete with wild salmon. If they are the same species as the wild salmon (i.e. Atlantic salmon grown in the Atlantic), there is the possibility of interbreeding between farmed and wild fish. When such interbreeding occurs, there is a significant change in the genetics of the salmon population. More.

Genetically engineered salmon: The introduction of genetically altered salmon in ocean pen farms — a distinct possibility in the near future — adds another layer of concern with respect to interactions with wild populations. If interbreeding were to occur as a result of escapes, such genes could be incorporated into the wild gene pool and possibly diminish the vigor of the wild population. Hatcheries are problematic because the selective survival of large numbers of young from the small numbers of adults that donate eggs and sperm change the genetic pool of the wild population. More.

Wild fish for feed: Farmed salmon are fed meal and oils from wild-caught fish. Each pound of salmon produced requires at least 3 pounds of wild-caught fish, challenging the presumption that fish farming necessarily reduces commercial fishing pressure. In fact, there is a net loss of protein in the marine ecosystem as a whole when wild catch is converted into meal for aquaculture consumption.

Pollution: Pens full of salmon produce large amounts of waste — both excrement and unconsumed feed. This may result in water quality conditions (such as high nutrients and low oxygen) that are unfavorable for both the farmed fish and the natural ecosystem. It is suspected that nutrients released from salmon farms stimulate micro- algal blooms, but proof of this is lacking because little research has been done.

Disease: The densely packed condition in pens promotes disease, a common problem in most salmon farms. Furthermore, there have been documented disease transfers from farmed salmon to wild populations, and the potential effects are serious. While antibiotics are used to treat some diseases, there are concerns about the effects of antibiotic-resistant bacteria on human health. There has been an emphasis on developing vaccines to prevent specific diseases, which reduces the need for antibiotics. More.

Aesthetics: In some areas, landowners have opposed salmon aquaculture facilities sited near residential shorelines because they are unsightly, odoriferous and/or interfere with the natural setting of the seascape. In fact, in the U.S., aquaculture development has been significantly slowed down by these concerns.

The Causes

Escapes are inevitable in open-water pens, especially during storm conditions. Some incidents have resulted in the escape of tens of thousands of farmed fish into the environment. Efforts to secure facilities against these accidents may reduce the size and number of releases, but is unlikely to stop them altogether.

The density of salmon in farms is variable, but the farmer is motivated to pack them at high densities to increase profits. This exacerbates the problems of pollution and disease and causes stress on the fish that leads to inferior product quality.

The siting of salmon farms is often problematic, particularly if it does not adequately take into account the proximity to wild salmon migration routes, water flow and circulation patterns, the fate of waste materials, the number of facilities already in an area and aesthetic concerns.

According to some, the development of genetically engineered salmon for farming is proceeding without concern for the consequences of inevitable escapes into coastal waters inhabited by wild salmon.

The Context

Populations of salmon in the western North Atlantic have recently plummeted for unknown reasons. Meanwhile, in the eastern Pacific, more than 100 populations have disappeared and salmon are extinct in 40 percent of the rivers where they once spawned along the North American Pacific coast. The potential for interactions with farmed fish and transmission of disease from farmed to wild salmon is especially threatening in the context of these declines.

Governments typically encourage aquaculture because it is viewed as economic development. This often leads to the intensive, large-scale farming methods most often associated with environmental damage.

The market price of farmed salmon is artificially low. Because the costs of environmental damage are not borne by the industry, nor are the value of ecosystem services factored into the cost of production, there is no pressure on the industry to operate in environmentally sound ways.

IMPACTS OF SALMONID FARMING ON MARINE COASTAL ECOSYSTEMS

The production of farmed salmonids in Chile reached 550 000 t in 2004. The industry is considered to be consolidated, but with potential for further expansion to the south into pristine coastal areas. The environmental impacts of the salmonid farming industry in Chile were reviewed in 1996, and evidence at that time did not suggest significant adverse effects. However, after almost ten years of sustained growth, current evidence indicates that significant loss of benthic biodiversity and localized changes in the physico-chemical properties of sediments have occurred in areas with salmonid farms. Furthermore, the presence of these farms significantly increases in pulses the density of dinoflagellates. Data suggest that escaped farmed fish may have an impact on native species, although their survival in the wild appears low. The abundance of omnivorous diving and carrion-feeding marine birds increased from twofold to fivefold in areas with salmon farms compared with control areas without them. It is urgent that an ecosystem approach be implemented to assess all impacts of salmonid farming on coastal ecosystems in southern Chile.

Introduction

During the past ten years, salmonid aquaculture has become the fourth largest economic activity in Chile after mining, forestry, and fruit production. Gross production in 2004 was 550 000 t, and the three most important farmed salmonids are Atlantic salmon, *Salmo salar*, coho salmon, *Oncorhynchus kisutch*, and rainbow trout, *O. mykiss* Farming of Atlantic salmon is the main aquaculture activity in Chile, accounting for 92.5% of all exported aquaculture biomass and 94.9% of the total export revenue derived from aquaculture products. The location of the salmon farming industry in Chile. Concern has been expressed about the environmental impacts of salmon farming in Chile and worldwide. These impacts include modification of benthic communities, increased nutrient loads in coastal waters and the associated problem of harmful algal blooms, increased harvests of wild fish populations for the production of fish feed, use of different types of chemicals, and escapes of farmed salmon into the wild (e.g. Gowen and Bradbury, 1987; Hallaegraeff, 1993; Folke *et al.*, 1998; Naylor *et al.*, 2000; Goldburg and Naylor, 2005). Currently, the environmental effects of intensive salmonid cultivation in Chile are little understood (Castilla *et al.*, 2006), and consequently, industry regulation has been criticized for failing to implement an integrated ecosystem approach to environmental conservation.

In 1996, a review of the environmental effects of salmon farming in Chile concluded that the industry did not have a significant environmental impact (Buschmann *et al.*, 1996). However, the industry has expanded considerably during the past ten years and is now considered to be economically and technologically consolidated. The industry may expand farther into pristine southern environments of high conservation value. This paper summarizes recent research on the most important environmental effects of salmon farming in Chile, based on published and unpublished information, and makes suggestions for the sustainable management of intensive salmon farming in the southeast Pacific.

Developments in Aalmon Farming in Chile

Commercial salmon farming began in Chile with the importation of eggs from the northern hemisphere. In recent years, the use of innovative technologies, particularly for Atlantic salmon culture, e.g. photoperiod and temperature manipulation, has permitted the transfer of fish to sea at different times of the year and harvesting throughout the year, reducing the need to import ova and, consequently, reducing the risk of introducing exotic parasites and diseases. Other positive developments include improvements in feed quality and conversion rates (ca. 1.2). Notwithstanding these improvements, several environmental issues remain unstudied or unregulated, resulting in pressure on producers and the government from environmental groups (Buschmann *et al.*, in press).

Impacts of salmonid farming on the Chilean coastal environment

Research conducted since 1996 suggests that there have been localized adverse impacts on the seabed in the licensed farming areas associated with physico-chemical changes to sediments and significant losses of benthic biodiversity (Buschmann, 2002; Soto and Norambuena, 2004). Research at eight salmon farm sites located along 300 km of coastline demonstrated that the benthic biodiversity was reduced by at least 50% on average in the licensed salmon farming areas in southern Chile. This loss of biodiversity appears related not only to organic matter loading and low oxygen levels in the sediments (Soto and Norambuena, 2004), but also to the deposition of copper (Buschmann, 2002). Recently, the channels and fjords of southern Chile have been shown to possess a unique benthic fauna comprising endemic cold-water corals, anemones, and other species (e.g. Försterra *et al.*, 2005). Interest in protecting these ecosystems has increased, and a network of conservation areas has been proposed (Castilla, 1996).

The Chilean regulations require that any farm causing the formation of anoxic sediments must reduce its production by 33% and, if the sediments have not recovered after one year, production must be reduced by a further 33%. Thereafter, if the level of oxygen in the sediments does not recover, the farm must close. However, benthic wastes have high levels of phosphorus (Soto and Norambuena, 2004), and it is not known how long these accumulations can persist and remain a source of nutrients in Chilean waters.

The occurrence of harmful algal blooms resulting from the input into coastal waters of nitrogen from salmonid farms is another major topic of concern in Chile. Harmful algal blooms have been recorded in Chile during the past three decades and have affected human health and natural and cultured marine resources. The first instances to affect human health were reported in 1970 and 1972, and involved the species *Dinophysis acuta* and *Alexandrium catenella*; in 1993, diatoms of the genus *Pseudo-nitzschia* were implicated in a harmful algal bloom. These algae are sources of diarrhoetic, paralytic, and amnesic shellfish poisons, respectively. *A. catenella* blooms have affected Chile's austral region (42 S–52 S), whereas *D. acuta* and *Pseudo-nitzschia* have resulted in blooms in various enclosed coastal areas. The algal species that have affected marine resources include Dinophyceae (*Prorocentrum micans*, *Gymnodinium* cf. *cloroforum*, and *Gymnodinium* spp.); Bacillariophyceae (*Leptocylindrus minimus*, *Chaetoceros convulutus*, and

Chatonella sp.); Raphydophyceae (*Heterosigma akashiwo*); and Chrysophyceae (*Dictyocha speculum*). These blooms have mainly affected the channels and fjords of the Chiloé region (41 S–43 S), causing behavioural changes and mortality in wild and cultured aquatic resources. Although the number of these harmful species appears to have declined in Chile, reports of new harmful species and bloom events have increased in the past three decades; the impacts of species in Chile that are recognized as being harmful in other parts of the world must be given careful consideration.

Harmful algae blooms in Chile during the past three decades, with the year of the bloom and its effects and details of potentially harmful algal species recorded in Chile.

In most locations, there have been no detectable increases in nitrogen concentration in the water column near salmon farms (Soto and Norambuena, 2004). However, in an intensively farmed channel in southern Chile (Calbuco), significantly higher concentrations of ammonium nitrogen were detected near the cages, compared with control areas without fish pens. Seaweeds were shown to exhibit faster growth rates and higher nitrogen content in their tissues when cultured near fish farm cages, suggesting higher nitrogen availability as a result of the presence of the farms (Troell *et al.*, 1997). In an experiment to assess the effect of salmon farms on phytoplankton communities, 1500-l tanks, filled with either effluent from salmon farms (see Buschmann *et al.*, 1994, for details of the experimental protocol) or with seawater pumped directly from the sea, were used to culture dinoflagellates. The results indicated that the density of dinoflagellates increased significantly ($p < 0.05$) when reared in fish farm effluent, while diatoms tended to disappear. In addition, a study using the "beyond BACI" (Before-After-Control-Impact) methodology (Underwood, 1994) suggested that salmon farms enhanced dinoflagellate abundance for short periods. An analysis of variance showed that no persistent effect could be detected on dinoflagellate abundance because the ratios of F-values T(Aft) I/T(Bef) I = 38.61 and T(Aft) I/T(Aft) C = 48.28 are both significant (p d" 0.05), whereas T(Aft) C/T(Bef) C = 0.24 (p e" 0.05) is not significant. These results indicate that the presence of salmon farms led to a significant increase in the abundance of dinoflagellates in short-term pulses and suggest that, to detect any increases in abundance of phytoplankton associated with them, it is necessary to implement an appropriate sampling protocol that facilitates detection of these pulses.

Ammonium concentrations (mg l"1) in seawater along a 20-km channel (Calbuco, 41 472 S 73 122 W) at sites with (+) and without (") salmon farms. Different letters denote significant (p d" 0.05) differences between sites after a Tukey's test. The same letter indicates no significant difference between sites.

Growth rate (top) and nitrogen content of tissue (bottom) for the red algae, *Gracilaria chilensis*, cultivated on suspended longlines at sites with salmon farms (Station A) and at control sites 1000 m away from the salmon farms (Station B). The experiment was replicated in two different seasons with the same results but of different intensity as a result of variability in local environmental conditions. Data from Troell *et al.* (1997). The figures present mean values and standard errors ($n = 10$).

Experimental induction of algal blooms in 1500-l outdoor tanks using water pumped directly from salmon rearing tanks and from the sea. (A) Phytoplankton cell density after a 15-d cultivation period ($n = 5$), (B) the proportion of the different taxonomic groups of algae present in tanks fed by salmon farm effluent, and (C) pumped seawater (AHB, unpublished data).

Changes in dinoflagellate abundance at a salmon farm site producing 100 t and at two control areas without salmon farms in the months before and after the installation (arrow) of the farm in Metri Bay (41 362 S 72 422 W). The sampling protocol involved the impacted site and two control areas located 1.5 km from the impacted site. Three samples were taken at the farm site and three in each control site during each sampling period. Each sample consisted of an integrated sample taken at 1, 6, and 12 m depth with a Nansen bottle. The data were analysed with SAS following the method of Underwood (1994).

ANOVA table for the detection of the effects of salmon farms on dinoflagellate abundance. The design allows for the detection of human impact when presenting a significant interaction between the control and the potentially impacted location and times and/or periods of sampling (see analysis after Underwood, 1994). The statistical analysis was performed with PROC GLM from SAS 6.12 for Windows. B = effect of the impact, T = effect of time, $T(B)$ = effect of time nested in impact, L = effect of the locality, BL = interaction between B and L, LT = interaction between L and T, and $LT(B)$ = interaction between L and $T(B)$.

The use of various chemical compounds in salmonid farming has also raised public concern in Chile, particularly following the detection of some compounds at high levels in other countries (Easton *et al.*, 2002; Hites *et al.*, 2004; Foran *et al.*, 2005). In this context, initiatives between the government and private companies in Chile have been set in motion. The effects of antibiotics on the microbial flora have received little attention, although resistance to antibiotics is a problem for the salmon farming industry, e.g. Miranda and Zemelman (2001). Statistics on the imports of antibiotics to Chile suggest their usage in large quantities in salmon farming (Cabello, 2003). It is necessary to reduce this usage to avoid a major environmental sanitary problem (Wolff, 2004). The environmental impacts of using malachite green, antifouling paints, and other chemicals have not been studied.

Escaped Farmed Salmon

The reported number of fish escaping from salmon farms in Chile varies annually, but the existing regulations have not led to a reduction in the reported biomass of escapees. Independent assessment of the validity of the reports is necessary. Farmed salmon released into the wild appear to have a rather short life expectancy (<1 y) in Chilean coastal waters (Soto *et al.*, 2001), as has been reported for the North Atlantic (Fleming *et al.*, 1996). Previous attempts at salmon ranching in Chile were not successful, owing to the low survival of the released fish.

Official reported biomass of salmonid escapees (t) in southern Chile.

There is concern in Chile, however, about the impact of escaped farmed salmon on the native fish fauna, although, because there are no natural populations of salmonids in Chile, escapees should not have the genetic impacts on indigenous populations reported in the North Atlantic (e.g. Fleming and Einum, 1997; Gross, 1998; Youngson and Verspoor, 1998; McGinnity *et al.*, 2003; Roberge *et al.*, 2006). Ecological interactions have been reported between farmed Atlantic salmon escapees and native Pacific salmon species (Volpe *et al.*, 2000), and understanding the ecological effects that escapees have on native fish communities in Chile is a research priority. Although limited studies of the prey of escapees have been undertaken, existing studies indicate that benthic invertebrates are the main prey items.

Commercial fishing for salmon is prohibited in Chile, although illegal fishing does occur. It has been proposed that fishing for salmon by craft fishers could be used to reduce the impact of escaped farmed fish (Soto *et al.*, 2001), although efforts to recapture escapees in the 1980s were unsuccessful. It has been suggested, however, that a legal fishery for escapees will encourage premeditated destruction of net pens to increase the availability of fish to support the fishery. The lack of clear evidence about the environmental impact of escapees and the absence of political support for a fishery targeting escapees (Buschmann *et al.*, in press) suggest that this management tool is unlikely to be introduced.

Other Impacts of Salmonid Farming

Salmonid farms can alter the natural transmission dynamics of sea lice to wild juvenile salmon with infestation pressures four orders of magnitude greater than the natural ambient level (Krkošek *et al.*, 2005). In Chile, preliminary data indicate that salmon farming can increase sea louse infestations on native fish populations (Sepúlveda *et al.*, 2004).

The abundance of marine birds is also affected by the presence of salmon farms, with the observed abundance of omnivorous diving and carrion-feeding birds increasing twofold to fivefold in some areas with salmon farms (66.2–77.2 birds h^{-1}) compared with control areas without them (13.8–31.0 birds h^{-1}). However, the top-down ecological consequences of these changes remain unstudied. In addition, the effects of salmon farms on marine mammals have not been evaluated. The South American sea lion, *Otaria flavescens*, frequently attacks salmon farms, although the frequency of attacks varies markedly and is not correlated with the proximity of the nearest sea lion colony (Sepúlveda and Oliva, 2005). It is important that the impacts of salmon farms on marine mammals be considered in future research.

Conclusions

Salmon farming is one of the most important economic activities in Chile, but its development has environmental impacts (Buschmann and Pizarro, 2001) that should be taken into account in regulating the industry. Further research is urgently required in Chile to increase understanding of these impacts, especially if the industry is to expand to the far south, a pristine coastal area that needs a sound conservation policy. It is important that a balance is maintained between further aquaculture development in Chile and environmental conservation through the development of integrated cultivation of extractive (i.e. macroalgae and filter-feeders) and fed organisms (fish) in the same waterbodies (e.g. Buschmann *et al.*, 2001; Chopin *et al.*, 2001; Troell *et al.*, 2003; Neori *et al.*, 2004). Ideally, Chile's low level of research capacity should be increased to improve current understanding of aquaculture–environment interactions.

GETTING A FRANKEN-LEG UP IN THE ANNUAL SALMON RUN

In the last Eco-Logical, we reported the findings of much higher levels of cancer-causing PCBs in farm-raised salmon than in wild salmon. That's reason enough to think about changing your behavior as a consumer of salmon, but just in case you're not quite convinced, let's wade around a bit more in the murky depths of salmon farming.

Other Chemicals

Beyond PCBs, farmed salmon contain higher levels of many other persistent chemical pollutants, including brominated compounds used as flame retardants and highly toxic pesticides such as DDT. Though PCBs are the most significant health concern related to farmed salmon, the other contaminants simply add to the overall concern about health risks associated with these fish.

Overuse of Fish-Meal Stocks

Fish farmers feed their farmed salmon fishmeal made from ground-up sardines, anchovies, mackerel, and other small fish that larger wild fish would normally eat. This limits the food supply for wild salmon, other predator fish like tuna, and non-fish species such as dolphin, which are already suffering declines due to overfishing of the smaller species they normally eat. Though it seems like fish farming should have an overall positive effect towards saving wild fish populations, it does not.

It should be noted that salmon-farming researchers are experimenting with fish feeds made from vegetable sources, but this approach is not yet a significant part of salmon farming operations, and it would still leave some of the other problems with the operations to be solved.

Water Pollution, Antibiotic Abuse, and Disease Transfer

The crowded conditions of a salmon farm often promote ill-health among the animals. Salmon farmers attempt to control diseases, parasites, and other problems by using antibiotics and pesticides. According to the Coastal Alliance for Aquaculture Reform, farmed salmon are fed more antibiotics per pound of body weight than any other livestock animal in North America.

Much of the antibiotic dosages given to farmed salmon pass through the fish unabsorbed. The excreted drugs settle in sediments and other parts of the environment near the fish farm, where they impact many of the smaller life forms that are part of the complex local ecosystem. Such effects have not been well evaluated. Overuse of antibiotics also leads to more virulent, drug-resistant strains of bacteria.

Salmon raised in farming operations located near wild salmon habitat can transfer diseases such as sea lice to the wild fish. Even though the nets don't allow the fish to intermingle, the free-swimming stage of the louse allows it to migrate from an infested farm area to nearby wild areas.

Fish feces builds up and pollutes the water and ocean floor in the area. Some salmon farmers also use anti-foulants on their net systems, and these chemicals—in addition to the pesticides and antibiotics—contaminate the water.

GM Salmon

Although farmed salmon are penned, they often escape. Since 1988, the industry has reported that over one million farmed salmon have escaped into Pacific waters, and because it's likely that some escapes have gone unreported, some experts believe that the real number of escapees is much higher.

If genetically modified (GM) salmon escape, it could have very bad consequences for wild salmon stocks. Although GM farmed salmon are less fertile than wild salmon, they mature earlier, are larger and thus more attractive to females, and are more aggressive at fertilizing eggs. Overall, they outpace wild salmon at the reproduction game. However, because these GM salmon have a much higher mortality rate in the wild, their dominance of the reproductive process could ultimately lead to the demise and extinction of the species.

9

ECOSYSTEM BASED FISHERIES MANAGEMENT

The traditional management strategy for fisheries and other living resources has been to focus on one species of fish and shellfish in isolation. For example, if there were a decline in the number of a certain kind of fish in the Bay, authorities might decide to decrease the number of that species that could be removed by fishing in a given year. But fishing of a single species is only one variable that affects the health of its population. Additional elements come in to play, such as interactions with other species and the effects of pollution and other stresses on habitat and water quality. To more effectively assess the health of any given fishery and to determine the best way to maintain it, the entire ecosystem must be taken into account.

An ecosystem is a geographically specified system of organisms (including humans), the environment, and the processes that control its dynamics. Ecosystem approaches to management use integrated approaches to study and manage the resources of an entire ecosystem. This approach considers the cumulative impacts from various sources and the balance of conflicting uses, and includes multiple factors such as pollution, coastal development, harvest pressure, predator/prey and other ecological interactions, and watershed management.

BACKGROUND

Aquatic ecosystems, including rivers, lakes and inland seas, flood plains, coastal lagoons and estuaries, coastal shelves and open oceans cover a very large part of the earth's surface and, among other amenities, goods and services, sustain the production of fisheries and aquaculture. They yield about 120 million tonnes of fish and fishery products per year - the largest source of wild protein - and provide a livelihood to as many as 140 million people. Fisheries and aquaculture exploit a large diversity or organisms ranging from algae, ascidians and sea-cucumbers to molluscs, crustaceans, fish and marine mammals.

Most aquatic ecosystems are unavoidably affected by fishery activities that involve a selective removal of part of the natural productivity for human subsistence, economic returns and development. However, undesirable fishing practices in some cases, such as overfishing and use of destructive methods, are unduly affecting these precious ecosystems, calling for urgent corrective action.

Except in the high seas, these ecosystems are also usually used for other purposes such as conservation (e.g. wetlands), forestry (e.g. mangroves), agriculture (e.g. floodplains), offshore mining and oil and gas extraction, and human settlements (e.g. coastal areas). Unfortunately, aquatic ecosystems are also, most often, the ultimate recipients of the pollution produced by human settlements and industrial activities, inland, on the coastal area as well as at sea. Even the most remote areas (e.g. deep ocean and polar seas) are now affected, seriously putting in question the sustainability of present practices and the present ecosystems resources to future generations.

All key international agreements adopted over the last two decades, including the 1995 FAO Code of Conduct for Responsible Fisheries, stress the need for the adoption of ecosystem approaches to fisheries (EAF). In response to these, in 2001, 57 countries issued the Reykjavik Declaration on Responsible Fisheries in the Marine Ecosystem which included a declaration of their intention to work on incorporating ecosystem considerations into fisheries management. The 2002 Plan of Implementation of the World Summit on Sustainable Development called for, amongst other things, the application of the Reykjavik Declaration by 2010 as one of the steps essential for ensuring the sustainable development of the oceans. Despite these good intentions and some progress in implementation in many parts of the world, in most if not all countries much progress in ecosystem research and institutional development is still needed before the implications of the approach are fully understood and credible management strategies are adopted and effectively implemented. In addition to governmental initiatives, environmental NGOs have been particularly active in raising awareness of governments and society and have proposed a number of basic principles for ecosystem conservation.

In the expression "Ecosystem Approach to Fisheries (EAF)" the terms ecosystem, approach, and fisheries are defined in dictionaries and in the scientific literature. Used together, however, they imply a process using specific means to achieve selected objectives. These objectives are often not explicitly defined in conventional fisheries management and are even more difficult to define clearly in EAF. Following on the Reykjavik Declaration, EAF is recognized as a form of fisheries governance framework, taking its conceptual principles and operational instruments from conventional fisheries management on the one hand, and ecosystem management on the other hand. A number of expressions related to fisheries and ecosystem management have been used to describe related concepts such as: ecosystem-based fisheries management (EBFM), ecosystem approach to fisheries (EAF), environmental management, biodiversity management and ecosystem approach (in the CBD). As modern fields of science-based governance, they all find their roots in the concept and well-grounded academic disciplines of natural resources management (NRM) or wildlife management but have evolved quite different operational paradigms. They also relate very closely to the already widely used concept of integrated management. These differences are briefly highlighted below.

Principles of ecosystem approaches to fisheries

The overarching principles of EAF are an extension of the conventional principles for sustainable fisheries development to cover the ecosystem as a whole. They aim to ensure that, despite variability, uncertainty and likely natural changes in the ecosystem, the capacity of the aquatic ecosystems to produce fish food, revenues, employment and, more generally, other essential services and livelihood, is maintained indefinitely for the benefit of the present and future generations. The FAO Technical Guidelines on the ecosystem approach to fisheries (FAO 2003) define EAF as follows:

"An ecosystem approach to fisheries strives to balance diverse societal objectives, by taking into account the knowledge and uncertainties about biotic, abiotic and human components of ecosystems and their interactions and applying an integrated approach to fisheries within ecologically meaningful boundaries."

A primary implication is the need to cater both for human as well as ecosystem well-being. This implies conservation of ecosystem structures, processes and interactions through sustainable use. Inevitably this will require considering a range of frequently conflicting objectives where the needed consensus may not be readily attained without equitable distribution of benefits. In general, the tools and techniques of EAF will remain the same as those used in traditional fisheries management, but they will need to be applied in a manner that addresses the wider interactions between fisheries and the whole ecosystem. For example, catch and effort quotas, or gear design and restrictions, will be based not just on sustainable use of the target resources, but on their impacts on and implications for the whole ecosystem.

The term has been defined (US National Research Council, 1998) as "an approach that takes major ecosystem components and services - both structural and functional - into account in managing fisheries... It values habitat, embraces a multispecies perspective, and is committed to understanding ecosystem processes... Its goal is to rebuild and sustain populations, species, biological communities and marine ecosystems at high levels of productivity and biological diversity so as not to jeopardize a wide range of goods and services from marine ecosystems while providing food, revenues and recreation for humans". The term puts the focus for management on the users. What is managed is the economic activity. The term did not meet with consensus at the 2001 FAO Reykjavik Conference, possibly because some countries took it as implying that the "ecosystem" would become the new "foundation" of fisheries management. This may have been interpreted as giving to environmental considerations pre-eminence over socio-economic and cultural ones, raising concern about equity, political as well as socio-economic costs and feasibility.

An ecosystem approach requires that fisheries management decisions consider the impact of the fishery not only on the target species, but also on non-target species, seafloor habitats, and the ecosystems of which these species are a part. This approach also requires that management decisions take into account changes in the ecosystem which may affect the species being fished. This includes the effects of weather and climate, and the interactions of target fish stocks with predators, competitors, and prey species.

Ecosystem-based fisheries management is also part of the broader approach to managing oceans and activities that take place in and around marine environments. Known asintegrated oceans management, it compels decision makers who are responsible for fisheries to consider other ocean users during management planning processes. This helps to ensure that fisheries managers make more informed decisions regarding ecosystem issues, such as the protection of ecologically significant areas and species.

Building on existing fisheries management practices, Canada's Sustainable Fisheries Framework forms a foundation for implementing an ecosystem approach in the management of its fisheries. This includes using new policies and tools to implement the precautionary approach to fisheries management decision making, and manage the impacts of these fisheries on sensitive benthic areas and forage species.

Elements needed for successful ecosystem management

Based on experience gained during 50 years of fisheries management, on the set of principles and points of operational guidance for ecosystem management recommended by the FAO Code of Conduct for Responsible Fisheries, the 5th Conference of the Convention on Biological Diversity (CBD), the 2001 Reykjavik Declaration, the FAO Technical Guidelines on EAF and on other instruments dealing with the subject, the following elements emerge as the foundations and components of an ecosystem management approach to fisheries and aquaculture:

- Recognize that management objectives are a matter of societal choice.
- Decentralize decision and action to the lowest appropriate level, while recognising that there must also be mechanisms to ensure that management decisions and actions are consistent and coordinated at the higher levels required by EAF.
- Identify the fishery or fisheries to be addressed in each case and the geographic area to be addressed. This will include matching fisheries management system boundaries with ecosystem boundaries.
- Introduce measures to ensure transparency, public awareness and consensus building.
- Establish effective conflict resolution and enforcement mechanisms.
- Ensure coordination, consultation and cooperation, including joint decision-making, between fisheries operating in the same ecosystem and other sectors that interact with it.
- Liaise with agencies and authorities responsible for non-fishery activities within the ecosystem to encourage and facilitate reduction of negative impacts on aquatic ecosystems from non-fishery activities.
- Recognize and identify the various direct and indirect uses and users of the ecosystem and involve all stakeholders in knowledge-sharing, decision-making and management.
- Translate the high-level policy goals for EAF into transparent and comprehensive operational objectives.
- Establish indicators and reference points for the agreed operational objectives in order to provide a framework for monitoring management performance.
- Introduce ecological accounting into fisheries management.
- Undertake action at the appropriate spatial and temporal scale, including setting management objectives for the short- and long term.
- Consider transboundary impacts of fisheries on adjacent or other ecosystems.
- Governance for EAF should ensure both human and ecosystem well-being and equitable allocation of benefits, as a condition for compliance.
- Understand and manage ecosystems in an economic context, including
  - (a) reduce those market distortions that adversely affect biological diversity,
  - (b) align incentives to promote biodiversity conservation and sustainable use, and

(c) internalize costs and benefits in the given ecosystem to the extent feasible.

- Establish appropriate, explicit and enforceable rights to ecosystems resources. Under EAF it needs to be recognised that access rights systems will frequently need to encompass other uses in addition to harvesting of target resources.
- Conserve ecosystem biodiversity, structure and functioning.
- Avoid irreversible ecosystem impacts from fisheries and reduce reversible, undesirable impacts to the minimum practically possible (e.g. bycatch and discards).
- Ensure an appropriate balance between conservation and responsible use.
- Recognize that ecosystem variability and changes are inevitable.
- Conservation and management decisions for fisheries should be based on the best scientific information available, also taking into account traditional knowledge of the resources and their habitat, as well as relevant environmental, economic and social factors;
- Recognize that while it is necessary to take immediate action to address particularly urgent problems, it is also important to advance the scientific basis for incorporating ecosystem considerations, building on existing and future available scientific knowledge. This could include improving knowledge on the structure, components and functioning of the marine ecosystem under consideration, the role of habitat and the biological, physical and oceanographic factors affecting ecosystem stability and resilience; improve the monitoring of by-catch and discards in all fisheries to obtain better knowledge of the amount of fish actually taken; support research and technological development of fishing gear and practices to improve gear selectivity and reduce adverse impacts of fishing practices on habitat and biological diversity; and assess adverse human impacts of non-fisheries activities on the marine environment as well as the consequences of these impacts for sustainable use.
- Under EAF, the application of a precautionary approach is particularly important because it is expected that uncertainty will be considerably greater than under traditional management focused on target resources only.

Fisheries Management

Modern fisheries management, as practised since the early 1940s, is strongly based on the ecosystem theory but focuses primarily on fishing activity and target fish resources. In inland waters, affected earlier and more strongly than marine waters by environmental problems, it developed as an extension of wildlife management (Lackey, 1999) and involves a substantial amount of direct intervention on the habitat, species composition, etc. In marine ecosystems, however, because the possibility of direct intervention on the ecosystem is limited, management strategies concentrated on controlling human intervention (fishing) while observing proxies for the state of an otherwise opaque ecosystem and fugitive resources. It is defined as "the integrated process of information gathering, analysis, planning, decision-making, allocation of resources and formulation and enforcement of fishery regulations by which the fisheries management authority controls the present and future behaviours of the interested parties in the fishery, in order to ensure the continued productivity of the living resources" (FAO, 1995b). It aims at optimizing the use of fishery resources as a source of human livelihood, food and recreation, dynamically regulating fishing activity, meeting resource-related objectives or constraints, mainly indirectly. It is science-based and has been evolving for 50 years with an increasing recognition of failures (Stevenson, 1973; Garcia 1996a; Goñi, 1998; Sutinen and Soboil, 2003).

Ecosystem Management

Ecosystem management, as a concept, has been formally around since at least the introduction of conservation ethics by Aldo Leopold in 1966 (in Czech, 1996, and Czech and Krausman, 1997). The concept emerged strongly at global level in the 1970s, initially boosted by the 1972 Stockholm Conference on the Human Environment and strengthened by the 1992 Conference on Environment and Development (UNCED) and Convention on Biological Diversity .

Ecosystem management derives from wildlife management, born on land, in range management and forestry. The latter involved direct manipulation of the habitat and population in space and age structure, as well as of human activity - in space, structure and time - with the view to optimizing long-term returns to humans (Lackey, 1998 and 1999). It reflects a stage in a continuing evolution of social values and priorities. It is area-based, and boundaries must be clearly and formally defined, notwithstanding the difficulty. It aims at maintaining ecosystems in the sustainable condition necessary to achieve desired social benefits. To be effective, it requires scientific information as an element in a decision-making process that is fundamentally one of public and private choice. There are plenty of overlapping definitions (Christensen et al., 1996) but a precise, universally accepted definition has yet to emerge (Lackey, 1999).

According to Lackey (1999), ecosystem management is also "the application of ecological, economic, and social information, options, and constraints to achieve desired social benefits within a defined geographic area and over a specified period". Ecosystem management can also be more comprehensively defined as: "a management philosophy which focuses on desired states rather than system outputs and which recognizes the need to protect or restore critical ecological components, functions and structures in order to sustain resources in perpetuity" (Cortner et al., 1994). It involves "management decisions which involve a broad awareness of the consequences of fishing or other human actions to an ecosystem....used to infer the necessity of understanding multispecies interactions and questions of altered structure of the biological community" (ecosystem stability)" (FAO-ACMRR, 1979). It aims at: (1) maintaining viable populations of all native species in situ; (2) representing within protected areas all native ecosystem types across their natural range; (3) maintaining evolutionary and ecological processes; (4) managing over periods of time of sufficient duration to maintain evolutionary potential of species and ecosystems; and (5) accommodating human use and occupancy within these constraints (Grumbine, 1994, cited by Larkin, 1996). The more recent literature tends to stress more the decision-making process (e.g. people's participation, transparency, adaptive management, precaution) than the elements to be conserved.

Ecosystem Approach

The term is usually used in the form of "ecosystem approach to..." as, for instance, in the ecosystem approach to fisheries (EAF) or to environmental protection (Gonzalez, 1996). Such an approach "recognizes explicitly the complexity of ecosystems and the interconnections among its component parts" (Fisheries and Oceans Canada, 2002). In general, the approach is taken as requiring: (1) definition and scientific description of the ecosystem in terms of scale, extent, structure, functioning; (2) assessment of its state in terms of health or integrity as defined by what is acceptable to society; (3) assessment of threats; and (4) maintenance, protection, mitigation, rehabilitation, etc., using (5) adaptive management strategies.

The 1992 UN Convention on Biological Diversity (CBD) refers simply to the "ecosystem approach" and defines it as "Ecosystem and natural habitats management... to meet human requirements to use natural resources, whilst maintaining the biological richness and ecological processes necessary to sustain the composition, structure and function of the habitats or ecosystems concerned. Important within this process is the setting of explicit goals and practices, regularly updated in the light of the results of monitoring and research activities".

It is also defined as "a strategy for the integrated management of land, water, and living resources that promotes conservation and sustainable use in an equitable way", as well as "a strategy ...to reach balance between... conservation, sustainable use, and fair and equitable sharing of the benefits arising from the utilization of genetic resources"(Fifth Meeting of the Conference of the Parties to the Convention on Biological Diversity (COP 5), Decision V/6).

It recognizes that "the ecosystem is a functional unit at any spatial scale... that humans are an integral part of many ecosystems... and requires adaptive management techniques." (Secretariat of the Convention on Biological Diversity, 2000). Its content is defined by the Twelve Malawi Guiding Principles (see Annex 1). The operational guidance is to: (1) focus on the functional relationships and processes within ecosystems; (2) enhance benefit sharing; (3) use adaptive management practices; (4) carry out management actions at the scale appropriate for the issue being addressed, with decentralization to the lowest level as appropriate, and (5) ensure intersectoral cooperation.

Ecosystem Approach to Fisheries (EAF)

The term "Ecosystem Approach to Fisheries" (EAF) was adopted by the FAO Technical Consultation on Ecosystem-based Fisheries Management held in Reykjavik from 16 to 19 September 2002 (FAO, 2003) for various reasons: (1) the reticence expressed by the Reykjavik Conference vis-à-vis the EBFM terminology (see next section); (2) the convenient parallel this term offers with the "Precautionary Approach" to fisheries; and, last but not least, the fact that the term EAF, not being limited narrowly to management, could easily cover also development, planning, food safety, etc., better matching the breadth of the FAO Code of Conduct. The term "approach" indicates that the concept delineates a way of taking ecosystem considerations into more conventional fisheries management, in line with the Reykjavik Conference wisdom. The EAF could be defined as the way, the spirit in which the Code ought to be implemented.

EAF is defined by Ward et al. (2002) as "an extension of conventional fisheries management recognizing more explicitly the interdependence between human well-being and ecosystem health and the need to maintain ecosystems productivity for present and future generations, e.g. conserving critical habitats, reducing pollution and degradation, minimizing waste, protecting endangered species". The Reykjavik FAO Expert Consultation (FAO, 2003) agreed that the "purpose of an ecosystem approach to fisheries is to plan, develop and manage fisheries in a manner that addresses the multiplicity of societal needs and desires, without jeopardizing the options for future generations to benefit from a full range of goods and services provided by marine ecosystems". Therefore, "an ecosystem approach to fisheries strives to balance diverse societal objectives, by taking account of the knowledge and uncertainties about biotic, abiotic and human components of ecosystems and their interactions and applying an integrated approach to fisheries within ecologically meaningful boundaries".

Despite the distinctions between ecosystem management and fisheries management described above, the underlying concepts remain still fuzzy and tend to overlap. The review above illustrates the fact that the various expressions refer to what appear to be in practice very converging, if not totally similar, processes, aiming at largely overlapping sets of objectives. All ecosystem-based approaches to management of economic activities "rely on similar precepts: the need for sound science, adaptation to changing conditions, partnerships with diverse stakeholders and organizations, and a long-term commitment to the welfare of both ecosystem and human societies" (Kimball, 2001). The various terminologies reflect the relative importance, explicit or not, given respectively to fisheries objectives and to ecosystem conservation in their narrow interpretation.

Integrated Management (IM)

The ecosystem-related concepts mentioned above have a lot in common and relate very closely to the already widely used concept of integrated management. The latter involves comprehensive planning and regulation of human activities towards a complex set of interacting objectives and aims at minimizing user conflicts while ensuring long-term sustainability. It implies the use of a collaborative/participative approach involving the main stakeholders in a flexible, responsible and transparent planning process, respectful of existing rights and duties. It recognizes the need to protect the ecosystem and the implications of multiple uses and aims at sustainable development. Taking account of uncertainty, it complies with the precautionary approach. It takes account of natural and economic areas and not only administrative or political ones. It specifically identifies ecosystem-oriented objectives and indicators. It acknowledges the fragmentation of the sectoral approaches and the linkages between inland, coastal and ocean uses. It integrates data collection, information and research (assessment) and recognizes traditional knowledge. It develops processes for stakeholders' interaction, particularly in objective setting, planning and implementation, including conflict resolution. It explicitly considers the cumulative effects of human activities, and its implementation is based on integrated management plans (IMPs) (see for instance Fisheries and Oceans Canada, 2002). This reading indicates clearly that EBFM and EAF are subsets or aliases of IM, possibly with a greater emphasis on the ecosystem implications.

INSTITUTIONAL FOUNDATIONS

The EAF, as interpreted by FAO (2003) does not represent a revolution (in the sense of a rupture with the current fisheries management paradigm). It is rather an important new phase in a process of continuous evolution of fisheries-related institutions. While the specific terms "ecosystem approach to fisheries (EAF)" or "ecosystem-based fisheries management (EBFM)" may not yet be common in international instruments, regional conventions or arrangements and national legislation, the underlying principles and conceptual objectives examined above appear in many of them (Kimball, 2001; Aqorau, 2003).

The institutional foundations of EAF are to be found in the numerous instruments and events which, during at least the last three decades, reflected a growing societal concern for our degrading environment. Selecting the most significant ones among them is partly subjective. Annex 2 contains the list of international events and instruments that we have selected as most relevant in paving the way to the ecosystem approach to management of the environment as well as of fisheries. These events and instruments materialize the progressive building up of institutional strength in parallel with progress in the understanding of the ecosystem functioning and of human institutions created to conserve or use them in a sustainable manner.

The two main roots of EAF are the UN Conference on Human Environment in 1972 and the UN Convention of the Law of the Sea adopted in 1982, two historical institutional processes from which the concept of sustainable fisheries development emerged. A number of international events have preceded, accompanied and followed the adoption of the 1982 Convention and UNCED and have contributed to the progressive emergence of the EAF paradigm. They are briefly examined below.

The FAO Technical Conference on Marine Pollution and its Effects on Living Resources and Fishing, Rome, 1970, was an early expression of the concern for land-based sources of pollution and degradation.

The UN Conference on Human Environment, Stockholm, 1972, dealt with the environmental aspects of natural resources management, stressing "the right of Man to modify the environment for its development and the dangers behind the huge capacity developed to do so". It stressed concepts central to the ecosystem management concept in general and to EAF in particular, such as people's participation, resource limitation, environmental degradation, demography, planning and management, institutions, role of science and technology, international collaboration and equity.

The FAO Technical Conference on Fishery Management and Development, Vancouver, 1972 (Stevenson, 1973), stressed both the problems of overfishing and of environmental degradation from non-fishery sources. It also called for new management approaches based on precaution and addressing multispecies problems. It proposed to design the new fisheries management into the broader framework of ocean management.

The 1980 Convention on the Conservation of Antarctic Marine Living Resources, which lead to the establishment of the Commission for the Conservation of Antarctic Marine Living Resources (CCAMLR), is usually considered a precursor of the ecosystem approach to fisheries. Its provisions require that any harvesting and associated activities must be conducted in accordance with the following principles of conservation (from Kimball, 2001):

1. A decrease in the size of any harvested population to levels below those which ensure its stable recruitment should be prevented. For this purposes its size should not be allowed to fall below a level close to that which ensures the greatest net annual increment.
2. The ecological relationships between harvested, dependent and related populations should be maintained, and depleted populations should be restored to the levels defined in 1 above.
3. Changes in the marine ecosystem which are not potentially reversible over two or three decades should be prevented or the risk of these changes should be minimized.

The evolution of the ecosystem approach in the CCAMLR context has been reviewed and analysed by Constable et al. (2000) in relation to the use of a precautionary approach, bycatch management, ecosystem impacts, etc. concluding that CCAMLR, which has not been able to conserve some of its most valuable resources, has still to face the real test in its ecosystem approach.

The 1982 UN Convention of the Law of the Sea formulated inter alia the basis for conventional fisheries management. It deals with maximum sustainable yield, the need for restoration of depleted populations, the interdependence of stocks (e.g. in Art. 61.3) and the issue of associated and dependent species (Art. 61.4 and 119.1.b). In addition, it stresses the obligation to protect and preserve the environment (Part XXII, Art. 192 and 193).

The World Commission on Environment and Development (WCED, 1984-1987) and the "Brundtland Report" (Our common future, WCED, 1987) further developed the concept of sustainable development. It stressed, inter alia, the concepts of intergenerational equity, sustainable use, prior environmental assessments, prior consultation, precaution and liability, cooperation on transboundary environmental problems and natural resources. The link between sustainable development and ecosystem-based management is illustrated by the definition of sustainable fishing adopted by the US Committee on Ecosystem Management for Sustainable Marine Fisheries, which defined it as "fishing activities that do not cause or lead to undesirable changes in biological and economic productivity, biological diversity, or ecosystem structure and functioning from one human generation to the next. Fishing is sustainable when it can be conducted over the long term at an acceptable level of biological and economic productivity without leading to ecological changes that foreclose options for future generations." (US National Research Council, 1998).

The related concept of "Ecologically Sustainable Development (ESD)" was adopted in the early 1990s in Australia in the wake of UNCED (Commonwealth of Australia, 1992) to emphasize the importance of the environment to long-term survival and to ensure that there was a balanced approach in dealing with environmental, social and economic issues. It was defined as "Using, conserving and enhancing the community's resources so that ecological processes, on which life depends, are maintained, and the total quality of life, now and in the future, can be increased". This definition emphasizes further the importance of the environment to long-term survival and the need for balance in dealing with environmental, social and economic issues. The ESD approach aims at three key objectives: (1) to enhance individual and community well-being and welfare by following a path of economic development that safeguards the welfare of future generations; (2) to provide for equity within and between generations; and (3) to protect biological diversity and maintain essential ecological processes and life-support systems.

The UN Conference on Environment and Development (UNCED, 1992) completed the work of the WCED, developing Agenda 21 as a basis for implementation. It led to the adoption of a number of conventions and agreements of relevance to EAF, such as the Framework Convention on Climate Change, the Biodiversity Convention and the UN Fish Stock Agreement. The Rio Declaration puts human beings "at the centre of concerns" (Principle 1) and recognizes the sovereign rights to exploit resources (Principle 2), as well as the responsibility to do so without damaging the environment beyond the EEZ (Principle 2). It recognizes, inter alia, the need to: cater for future generations (Principle 3); integrate environmental protection in development (Principle 4); eliminate unsustainable patterns of production and consumption (Principle 8); encourage public participation (Principle 10); widely apply the precautionary approach, internalize environmental costs (Principle 16); the need to apply the Polluter Pays principle and undertake environmental impact assessment (Principle 17); the role of women (Principle 20) and indigenous communities (Principle 22); and the need for peaceful conflict resolution (Principle 26).

The 1992 UNCED Agenda 21 takes an ecosystem approach to ocean management. Its Chapter 17 calls for "new approaches to marine and coastal area management and development which is ... integrated in content... precautionary and anticipatory in ambit". It recognizes that use of marine resources and environmental protection are inseparable, and that integrated management is necessary to protect both. It addresses in detail the integrated management and sustainable development of coastal areas (Programme A), marine environmental protection (Programme B), sustainable use and conservation of marine living resources in the high seas (Programme C) and in areas under national jurisdiction (Programme D). It also addresses uncertainties related to natural variability of the marine environment and climate change (Programme E). Programmes C and D are particularly relevant for fisheries. They provide for, inter alia, strengthening of conventional management (to eliminate overfishing) as well as multispecies management, associated and dependent species, relations between populations, restoration of depleted stocks, improvement of selectivity and reduction of discards, protection of endangered species and habitats, prohibition of destructive fishing, and the role of science.

The 1992 Convention on Biological Diversity (CBD) elaborates the core principles of multiple-use management of biodiversity. It emphasizes the conservation of biodiversity, the sustainable use of its components and the fair and equitable sharing of benefits. Under the Convention, parties have the right to exploit and use biological resources and have an obligation to manage activities that may threaten biodiversity, regardless of where those effects may occur, and to collaborate where these effects occur on the high seas. The CBD complements and reinforces the 1982 UN Convention on the Law of the Sea (Kimball, 2001). The CBD elaborates also on the 1982 Convention's content in relation to genetic resources and genetically-modified organisms (GMOs). Furthermore, the CBD identifies the establishment of a system of marine protected areas as a key measure for conservation of biodiversity. The CBD definition of Biodiversity includes ecosystem diversity (the variety and frequency of occurrence of different ecosystems), species diversity (the frequency of occurrence of different species) and genetic diversity (the frequency of occurrence and diversity of different genes and/or genomes within species). Biodiversity is important from an EAF point of view because it is related to "resilience" or capacity to resist an impact or return to original conditions after the impact is removed. As a consequence, it is of interest to fisheries that the diversity of exploited habitats and the diversity of habitats and species in them is maintained and possibly enhanced as an "insurance" against negative consequences of future changes.

The 1995 Jakarta Mandate on Marine and Coastal Biological Diversity elaborated further on the "ecosystem approach" adopted by the CBD, further specifying it and basing it on: protected areas; the precautionary approach; scientific knowledge; indigenous knowledge and stakeholders' participation. It aims, inter alia, at integrated management; development of the ecosystem approach; valuation and effects of marine protected areas; assessment and minimization of mariculture impacts and the understanding of causes and impacts of the introduction of alien species (Pirot et al., 2000; Kimball, 2001).

The 1995 UN Fish Stock Agreement (FSA) aims at long-term conservation and sustainable use of these marine living resources, recognizing from the onset "the need to avoid adverse impacts on the marine environment, preserve biodiversity, maintain the integrity of marine ecosystems and minimize the risk of long-term or irreversible effects of fishing operations". The FSA deals with the precautionary approach, protection of biodiversity and sustainable use of fisheries resources. It calls on participating states to, inter alia: (1) protect biodiversity in the marine environment; (2) adopt measures to ensure the long-term sustainability of the fish stocks and promote their optimum utilization; (3) take account of environmental and economic factors; (4) adopt an ecosystems approach, whereby dependent or associated species are taken into account; and (5) take measures to prevent or eliminate overfishing and excess fishing capacity. It also provides for the precautionary approach, and the way of applying it is detailed for the first time, as through the specification of precautionary reference points and identification of management actions to be triggered in relation to them. It finally promotes aprinciple of compatibility, according to which management measures taken in different jurisdictional areas must be compatible across the entire area of distribution of the stocks.

The 1995 FAO Code of Conduct for Responsible Fisheries provided a voluntary framework to increase the sustainable contribution of fisheries to development. It combines the provisions of the 1982 Convention, the 1995 Fish Stock Agreement and the CBD and, as such, is the generally recognized global framework for capture fisheries and aquaculture, in marine or inland waters, in EEZs or the high seas. Its provisions are examined in detail further below.

The 1995 Kyoto Declaration on the Sustainable Contribution of Fisheries to Food Security emphasizes the importance of fisheries as a food source for the world's population. It sets out a number of principles that focus on sustainable development of fishery resources related to maintaining food security. It contains the agreement to undertake immediate action to, inter alia: "conduct.... integrated assessments of fisheries in order to evaluate opportunities and strengthen the scientific basis for multispecies and ecosystem management..." and to "minimize post-harvest losses...".

The 2001 Reykjavik Declaration on Responsible Fisheries in the Marine Ecosystem addressed directly and specifically the issue of introducing more ecosystem considerations into conventional fisheries management. Referring to the 1982 Convention, UNCED and the Code of Conduct, it recognized the need to "take into account the impacts of fisheries on the marine ecosystem and the impacts of the marine ecosystem on fisheries", and confirmed "that the objective of including ecosystem considerations in fisheries management is to contribute to long-term food security and to human development and to assure the effective conservation and sustainable use of the ecosystem and its resources." It recognized "the complex interrelationship between fisheries and other components of the marine ecosystems" but stressed that "including ecosystem considerations in fisheries management.... would enhance management performance". It called for "incorporation of ecosystem considerations such as predator-prey relationships" and for a better "understanding of the impact of human activities on the ecosystem". It emphasized "the role of science and the impact of non-fishery (usually land-based) activities".

The Reykjavik Declaration called for, inter alia:

(1) immediate introduction of management plans with incentives for sustainable use of ecosystems;
(2) strengthening of governance;
(3) prevention of adverse effects of non-fisheries activities on the marine ecosystems and fisheries;
(4) advances in the scientific basis for incorporating ecosystem considerations in management (including the precautionary approach);
(5) monitoring of interactions between fisheries and aquaculture;
(6) strengthening of international collaboration;
(7) technology transfer;
(8) removal of trade distortions; and
(9) collection of information on management regimes.

While calling generally for the continued implementation of Agenda 21 (in its Article 29b), the Plan of Implementation of the World Summit on Sustainable Development(WSSD, Johannesburg, South Africa, September 2002) in Paragraph 29d of the advance unedited Draft Plan of Implementation for the World Summit on Sustainable Development reinforced the conclusions of the Reykjavik Declaration by encouraging "the application by 2010 of the ecosystem approach, noting the Reykjavik Declaration on

Responsible Fisheries in the Marine Ecosystem ....". It also provides for "the use of diverse approach and tools, including the ecosystem approach, the elimination of destructive fishing practices, the establishment of marine protected areas...".

While it cannot easily be related to one particular international meeting, the recent and progressive development of ecolabelling in fisheries aims at establishing higher congruence between trade and sustainability objectives (Deere, 1999). Its application to fisheries has recently attracted a lot of international attention following UNCED. The potential usefulness of ecolabelling schemes to create market-based incentives for environmentally friendly products and production processes was internationally recognized at UNCED, where governments agreed to "encourage expansion of environmental labelling and other environmentally related product information programmes designed to assist consumers to make informed choices". The Marine Stewardship Council (MSC) has developed an ecolabelling system for capture fisheries, and the Global Aquaculture Alliance (GAA) is developing one for aquaculture products. Discussions are also ongoing regarding the possibility of applying to fisheries more generic systems such as ISO. While the future of the concept is still far from clear, fisheries managers are now aware that the future value of their fisheries might well be soon related to their management performance in terms of impact on the ecosystem.

Designing tools to evaluate fishery management strategies

Ecosystem objectives in fisheries management usually flow from high-level national policies or strategies and international agreements. Consequently they are often broadly stated and hence are difficult to incorporate directly in management plans. Predicting the results of any management action is very uncertain because the dynamics of ecosystems are complex and poorly understood. Methods to design and evaluate operational management strategies have advanced considerably in the past decade. These management-strategy-evaluation (MSE) methods rely on simulation testing of the whole management process using performance measures derived from operational objectives. The MSE approach involves selecting (operational) management objectives, specifying performance measures, specifying alternative management strategies, and evaluating these using simulation. The MSE framework emphasizes the identification and modelling of uncertainties, and propagates these through to their effects on the performance measures. The framework is outlined and illustrated by three ecosystem-related applications: management of benthic habitats and broad fish community composition; by-catch of species of high conservation value; and food-chain interactions and dependencies.

Techniques for quantitative fishery management have evolved rapidly during a period when computers, programming languages, and computational algorithms have also changed dramatically. Despite these advances, many stock assessment methods remain untested. A process of management strategy evaluation (MSE) could potentially rectify this problem, but it would require a framework in which to conduct systematic tests. We survey the tools currently used for stock assessments and discuss the development of new standards for testing management procedures. A successful project would depend on human skills scattered among various nations, organizations, and academic disciplines. Analogies from civil engineering illustrate the discipline and collaboration required for an effective outcome. If the world community of fishery scientists could design, build, and support such a project, it would revolutionize the theory, teaching, and practice of scientific fishery management.

Fisheries stock assessment methodology has progressed rapidly during the past three decades. Most of this progress would have been impossible without corresponding improvements in the tools available for computation. For example, the modern technique of management strategy evaluationrequires extensive simulations to test the effectiveness of proposed management strategies. Obviously, this depends on an ability to perform simulations rapidly and efficiently.

As stock assessment science matures, the computational requirements will continue to increase, but fast hardware and software alone do not guarantee progress. To be really effective, the world's research community needs to adopt standards that facilitate cumulative knowledge, in which clear guidelines emerge for the conduct of stock assessments and their practical application to management. We discuss the current situation and offer suggestions for the future. We also speculate on the prospects for a computational framework that would meet the needs of stock assessment scientists worldwide. In particular, we examine the implications of this idea for designing the next generation of stock assessment software.

ECOSYSTEM APPROACH FISHERIES PRINCIPLES

The EAF amounts to introducing a series of modifications to conventional fisheries governance with the view to improving its poor performance (Sutinen and Soboil, 2003). As such, it is a fundamental contribution to sustainable development and is underpinned by all its principles. It follows that EAF is also a more comprehensive approach to the implementation of the Code of Conduct for Responsible Fisheries in all its aspects, from assessment to management and from capture to processing and trade, and that the general principles of the Code are fully relevant to EAF.

The various forms of ecosystem management or ecosystem-based management described in the literature or adopted formally by states (e.g. under the Convention on Biological Diversity framework) refer to a number of interrelated guiding principles or conceptual objectives, many of which are fairly generally agreed to. Some of them were established formally in the 1982 Convention on the Law of the Sea. Others are derived or extended from that convention. They are reviewed in the following sections, the order of which does not imply priority.

Human and Ecosystem Well-being

- Human well-being: A condition in which all members of society are able to determine and meet their needs and have a large range of choices to meet their potential.
- Ecosystem well-being: A condition in which the ecosystem maintains its diversity and quality - and thus its capacity to support people and the rest of life - and its potential to adapt to change and provide a viable range of choices and opportunities for the future. (Prescott-Allen, 2001).

The ecosystem approach to fisheries (EAF) is also a strengthened approach towards sustainable development of fisheries, recognizing more explicitly the interdependence between human well-being (HWB) and ecosystem well-being (EWB). It recognizes the need to maintain the productivity of ecosystems for present and future generations, conserving critical habitats, reducing pollution and degradation, minimizing waste and protecting endangered species. It also recognizes that this will not be achieved without the cooperation of people, i.e. unless the ecosystem contributes to human well-being, providing sustainable goods and services and sources of livelihood.

Resource Scarcity

The fact that aquatic ecosystems' resources are finite and often in short supply has now become an axiom. The resources are renewable but the quantities that can be extracted every year without jeopardizing the renewal capacity are limited and, in many cases, insufficient to satisfy the potential demand. The consequence is that many of them are depleted, leading to severe social and economic disruptions. A related objective is to assess the limits of the resources and the conditions for its maintenance (research), to regulate the extractive capacity of the fishery to match removals and to maintain critical ecosystem process and structures.

Maximum Acceptable Fishing Level

Article 61.2 of the 1982 Convention on the Law of the Sea provides that "states should ensure that the maintenance of the living resources in the exclusive economic zone is not endangered by overexploitation". This principle is reflected in the agreement establishing regional fishery management bodies and in most national fishery legislations. For instance, the Australian ESD charter provides that "a fishery must be conducted in a manner that does not lead to overfishing". While overfishing is not always precisely defined, the related objective is to allow catch levels (or fleet sizes) compatible with the maintenance of an ecologically viable stock at an agreed level, or range of levels, with acceptable probability that the set-up is viable. In this respect, the control of fishing activity, e.g. in terms of fishing capacity, gear and practices, is central to a responsible ecosystem approach to fisheries development and management.

Maximum Biological Productivity

"Exploited populations must not be allowed to fall below a level close to that which ensures their greatest net annual increase" (1980 Convention on the Conservation of Antarctic Marine Living Resources). This objective may be considered as an older formulation of the preceding one (on overexploitation) in which the maximum acceptable fishing level is that corresponding to the Maximum Sustainable Yield (MSY) at which the annual natural rate of increase is greatest. This principle has been central to conventional management as established in the 1982 Convention (Article 62.3), which provides that "measures shall also be designed to maintain... populations of harvested species at levels which can produce the maximum sustainable yield, as qualified by relevant environmental and economic factors". As above, the related objective is to allow catch levels (or fleet sizes) compatible with the maintenance of a stock at or above the MSY level. Indeed the UN Fish Stock Agreement has established that, for precautionary purposes, MSY should be considered as a "limit" to be avoided and not as a target to be reached.

Impact Reversibility

The 1982 Convention (Article 62.3) requires States to "restore populations of harvested species at levels which can produce the maximum sustainable yield". This principle is also reflected in the Australian ESD charter which provides that "for those stocks that are accidentally overfished, the fishery must be conducted such that there is a high degree of probability that the stock(s) will recover". The same principle has also been adopted in relation to the ecosystem and provides that "risks of changes to the marine ecosystem that are not potentially reversible over two or three decades must be minimized". The US National Marine Fisheries Service Panel on EBFM also noted as a principle that "once (ecosystem's) thresholds and limits have been exceeded, changes can be irreversible" (Fluharty and Cyr, 2001).

Impact Minimization

This objective complements the two above as keeping impacts at the lowest possible level may be an effective way to ensure the highest probability of reversibility. The principle provides that "fishing operations should be managed to minimize their impact on the structure, productivity, function and biological diversity of the ecosystem". The principle is reflected in Article 5f of the UN Fish Stock Agreement. Related objectives are to conduct fisheries in a manner that (i) does not threaten bycatch species; (ii) avoids mortality of, or injuries to, endangered, threatened or protected species and (iii) minimizes the impact of fishing operations on the ecosystem generally. The latter is the most difficult to implement as it is open-ended, in the absence of clear socio-economic objectives to be met while "minimizing" the impact.

Rebuilding of Resources

When stocks are overfished, depleted populations must be restored to stable recruitment levels (1980 Convention on the Conservation of Antarctic Marine Living Resources). The principle is also reflected in the 1982 Convention on the Law of the Sea, Article 62.3, which requires that "measures shall also be designed to ...restore populations of harvested species at levels which can produce the maximum sustainable yield, as qualified by relevant environmental and economic factors". The related objective is to plan for and implement within mandatory time frames a rebuilding strategy for exploited stock(s) which are below the agreed (preferably precautionary) reference points.

Ecosystem Integrity

It is often stated that the "integrity" of the ecosystem should be maintained, preserved and in any case aimed at. While there is apparently no widely agreed definition of integrity, the application of the principle is taken, in the CBD, as implying or requiring: (i) maintenance of biodiversity at biological community, habitat, species and genetic levels and (ii) maintenance of the ecological processes that support both biodiversity and resource productivity.

Species Interdependence

The 1982 Convention refers to the need to "take account of ... the interdependence of stocks" (Article 62.3) and provides that the "coastal

State shall take into consideration the effects on species associated with or dependent upon harvested species with a view to maintaining or restoring populations of such associated or dependent species above levels at which their reproduction may become seriously threatened" (Article 62.4). The principle is reflected in Article 5b of the UN Fish Stock Agreement. The 1980 Convention on the Conservation of Antarctic Marine Living Resources provides that "ecological relationships between harvested, dependent and related species must be maintained". This principle often refers specifically to endangered, threatened or protected species (see above). The related objective is to minimize bycatch and discards. As it is impossible to optimize the exploitation for all species at the same time, compromise solutions will need to be found, reflecting decisions on which species may be more negatively affected.

Institutional Integration

"States should ensure that an appropriate policy, legal and institutional framework is adopted to achieve the sustainable and integrated use of the resources, taking into account the fragility of coastal ecosystems and the finite nature of their natural resources and the needs of coastal communities" (The Code, Article 10.1). An expression of this principle can be found also in the WWF Guidelines (Ward et al., 2002) which state that "ecosystems are of value to society and can potentially be used in many ways, to satisfy various sectors' needs and strategic interests, now and in the future". The related objective would be to ensure that fisheries management takes account of interactions with other types of uses of the ecosystem. This implies a need to develop functional connections between fisheries management institutions, other sectoral institutions, and other institutions in charge of ecosystem maintenance.

Uncertainty, Risk and Precaution

Aquatic ecosystems are poorly known. They are complex, dynamic, changing seasonally and in the longer term and modified by fisheries, aquaculture and other activities. They are interconnected, potentially leading to significant transboundary effects. As a consequence, the resilience of ecosystems and the extent and reversibility of human impacts are difficult to forecast and hard to distinguish from natural changes. The related objectives are to: (1) improve research to better understand ecosystems; (2) take measures that account for complexity and dynamics and are robust to uncertainty, and (3) give attention to transboundary impacts. These objectives are encapsulated in the Precautionary Approach.

Compatibility of Management Measures

Boundaries of ecosystems and of present jurisdictions are unlikely to be fully compatible. In particular, many ecosystems will straddle political boundaries between EEZs or extend in the high seas. The implication is that management measures need to be coherent across the resource range. The UN Fish Stock Agreement is innovative in this regard and provides that "conservation and management measures established for the high seas and those adopted for areas under national jurisdiction shall be compatible in order to ensure conservation and management of the straddling fish stocks and highly migratory fish stocks in their entirety" (Article 6.2). The related objective is to promote collaboration between subnational or national authorities (as relevant) to ensure that measures taken under different jurisdictions converge towards common objectives. This also implies that objectives are indeed to be agreed on.

The Polluter Pays Principle (PPP)

The Polluter Pays Principle, which is well established in the literature related to pollution and sustainable development, provides that "the polluter should bear the cost of the measures needed to ensure that the ecosystem is and remains in an acceptable state" (Dommen, 1993). Such measures would include those aiming at prevention and control of pollution as well as mitigation of impacts and rehabilitation of affected areas. This principle is of direct relevance to fisheries where the sector is a source of significant pollution (e.g. fish processing) as well as dumping or involuntary loss of fishing gear. It could perhaps be extended to cover habitat damage. The principle can be implemented by various means, ranging from establishing process and products standards to application of taxes.

The User Pays Principle (UPP)

The User Pays Principle is also abundantly referred to in the literature about sustainable use and, as the Polluter Pays Principle, also aims at fuller internalization of production costs. It states that "all resource users should pay for the full long-term marginal social cost of the use of a resources and related services including any associated treatment cost" (Dommen, 1993, p.151). In other terms, authorized users should pay for the exclusive privilege granted to them to use a public resource. The principle can be implemented through payments for licences or quotas or though taxes.

The Precautionary Principle and Precautionary Approach

The Precautionary Principle and the way to implement it (the precautionary approach) are embedded in the UNCED Declaration

(Principle 15) which provides that "the precautionary approach should be widely applied and that, where there are threats of serious or irreversible damage, lack of full scientific certainty should not be used as a reason for postponing cost-effective measures to prevent environmental degradation". The approach has been adopted for fisheries in the UN Fish Stock Agreement and the FAO Code of Conduct, and guidelines are available for its practical implementation (FAO, 1996a). Such implementation has already started in some areas, mainly in the developed world (Garcia, 2000), but the available experience is too limited yet to allow for an appraisal of its outcome.

Subsidiarity, Decentralization and Participation

Subsidiarity is a concept of governance in which "decisions should be taken at the lowest possible level" (Bothe, 1998). Together with decentralization and devolution, subsidiarity is more and more often invoked with the view to increasing direct involvement of stakeholders in decision-making. The concept is obviously attractive, considering the relative failure of central government systems, and is supported by the general trend towards reduced governmental power, characteristic of globalization. The process implies the creation of institutions at lower governance levels and the development of governance capacity at such levels. Such institutions and processes existed in traditional fisheries in all continents but have progressively disappeared, except in some areas, destabilized by modernization and intrusion of the market economy. However, the concept which appears successful in developed countries (where it is seen as an extension of the regionalization process) may be difficult to implement in low-capacity areas and in freshly independent countries in which nationhood is a recent and still fragile concept.

Contributing to subsidiarity, decentralization is intended to increase participation of lower-level governance to decision-making. Participation may be implemented at many different degrees of involvement of the stakeholders in the management process in data collection, knowledge-building, option analysis, decision-making and even implementation, including enforcement. It calls for a decision-making framework that meaningfully includes and takes account of all sectoral and community interests. This is achieved by active participation from stakeholders, capacity building at appropriate level, widest possible distribution of information and dispute resolution mechanisms (e.g. at the lowest possible level, if the subsidiarity principle is applied).

Equity

Governance should endeavour to establish and preserve equity in all its forms: intergenerational, intragenerational, cross-sectoral, cross-boundary and cross-cultural, with special attention given to rights of minorities. Equity implies that similar options are available to all parties, a principle of stewardship by governments and the community. A number of sub-concepts have been referred to but may not meet with consensus. Intergenerational equity, for instance, is widely referred to and requires that future generations be given the same opportunity as the present ones to decide on how to use the resources. It can be sought through avoidance of actions that are not potentially reversible on some agreed time scale (e.g. a human generation), consideration of long-term consequences in decision-making and rehabilitation of degraded physical and biological environments. Lack of intragenerational equity (i.e., equity among sections of the present generation) is recognized as one major source of conflict and source of non-compliance. Intersectoral equity would seem very hard to define and operationalize but would imply, for instance, that the fishery sector be fairly treated when its interests conflict with those of other sectors. Cross-boundary equity may be a condition to successful shared stocks agreements. Intercultural equity is relevant when allocating resources to different cultures or defining rights of minorities (e.g. between indigenous and other populations).

IS ECOSYSTEM APPROACH A FUSION OR COLLISION?

Evolution Versus Revolution

EAF is an evolution of the fisheries management paradigm which borrows some central principles of ecosystem management and gives to them a practical operational meaning. However, both paradigms are evolving and the interaction between their respective trajectories is not immediately obvious. While this paper could not address the evolution of ecosystem management (sensu lato), it underlines the fact that ecosystem management and fisheries management have become two related but fairly distinct paradigms, with their independent schools of thought, scientific guilds, line ministries, international instruments and regional institutions. It also underlines the fact that they are evolving and converging, exchanging principles and elements of action. Their decades-long evolution, on parallel if not diverging routes, is now inflected towards a closer encounter or violent collision, depending on areas, governance systems and types of fisheries. We argue that a violent collision would be detrimental to society, environment and fisheries.

Co-evolution of Science and Governance

The process is one of co-evolution of governance and science, a symbiosis through which, supporting and criticizing each other, they were both forced to evolve (Garcia, 1996a) and become more conscious of: (1) the long-term costs of short-term "fixes"; (2) natural variability, uncertainty and need for precaution; (3) the need for more democratic, participative and transparent processes, as well as more powerful oversight mechanisms; (4) the importance of cooperation, coordination and integration. However, the fundamental perceptions remain still at some angle. While the long-term goals of the two processes are converging, there are sources of friction, e.g. in the attention to be given to social and economic stress in decision-making and in selecting options, as well as in the expectations about the acceptable time horizon for implementation.

Parallel Evolution of Paradigms

Ecosystem management, on the one hand, has developed and maintained a body of (largely qualitative) knowledge on ecosystem structure and functioning. It is developing sets of sustainability indicators. It has tested management organizations, processes and instruments (e.g. protected areas) and rediscovered that people, their aspirations, costs and benefits, allocation and equity - with the potential conflicts they can generate - have to be taken into account.

Fisheries management, on the other hand, has developed a body of (more quantitative) knowledge on population dynamics, interactions between fisheries and target resources, consideration of uncertainties in assessment and management and sustainability indicators. It has tested management institutions at all levels, has experienced 20 years of space allocation following the adoption of the 1982 Convention and rediscovered that maintaining associated and dependent species, as well as critical habitats and ecosystem processes, is necessary.

In both processes, the players have discovered that conflicts must be tackled and resolved equitably, that enforcement capacity must be improved and both are testing the assumption according to which increased participation, decentralization and transparency will improve management performance. With time, through scientific progress, recurrent failures and growing pressure from developing environmental ethics, both paradigms are evolving towards a more balanced approach to ecosystem and human well-being. It is recognized that both needs are intricately interconnected and that conflicts between users' requirements need to be addressed and resolved.

The shift to new terminology reflects this evolution. Fisheries management has started to take on more environmental and biodiversity considerations, as shown by the evolution of ecosystem-related provisions from the 1982 Convention to the UN Fish Stock Agreement and the FAO Code of Conduct in 1995. In parallel, ecosystem management has started to take on more socio-economic and cultural considerations, progressively accepting humans as a necessary part of the ecosystem and the satisfaction of their needs as a condition for sustainability and conservation. The examination of both trajectories indicates that, while probably starting from a common concern and set of principles far in human history, the two management paradigms have developed different (albeit convergent) focus and objectives, based on different perspectives, processes and institutions. They both met with dismal performance and rediscovered that the needs and constraints of both the people and the environment must be satisfied.

Obstacles to the "Fusion"?

It would seem obvious that there is a need to join the paradigms as they probably were centuries ago in more primitive civilizations and as attempted in the concept of sustainable development, but there are some difficulties on the way:

- The two paradigms are underpinned by two separate families of institutions, organizations and legal instruments and two types of scientists similar by their academic background but often different by the institutional system in which they operate. This creates the costly need for coordination and collaboration but also the political incentives to maintain separate identities with related fights for power and budgets. In other words, the institutions created as part of the solution may well become part of the problem across the way towards a joint (fused) paradigm.
- The two processes are somewhat different. Fisheries management decisions are based on evolving science (at least in theory), the outcome of which is considered (and sometimes applied) through an administrative process that has shown strong inertia, reinforced by the inherent resistance to change of the sector, struggling with the direct practical, social, economic and political consequences of the change. Fisheries management has been intensively tested in all contexts. It has been used for decades and its strong points and weaknesses are well known. Ecosystem management is also supported by science but decision processes (particularly at international level) seem to operate under higher public pressure often organized by NGOs through an efficient use of the media. It is evolving rapidly, supported by a large number of citizens, most of them with limited or no understanding of the costs of change to the sector and who often assume a zero cost to themselves. It has had a much more limited time for operational application and has accumulated little experience in accounting for social and economic issues. The difference in history and processes makes it more difficult for the two paradigms to "merge".

Both paradigms, however, must confront the difficult and recurrent question of allocation of resources and wealth at societal, cross-sectoral and even individual levels, the complexity of the "equity" question and the potential for conflict among a large and diversified group of stakeholders.

Outlook

It is not easy to forecast how the evolution of the two paradigms will proceed. Such evolution will probably depend partly on the evolution in other sectors than fisheries (e.g. water, forestry, waste, etc.). In any case, fisheries will most probably be considered from now on as one "environmentally degrading" industry among many others, a fact that the sector must take seriously into account. During the next two decades, the process could evolve as:

(1) A "violent" collision between radicalized paradigms, thrown out of balance by globalization, conflicts between stakeholders, and no reversal in overfishing and pollution trends. In this scenario, public pressure increases on a sector perceived as adapting too slowly and unwillingly to new societal requirements. Considering the respective political and electoral weights of the two stakeholder groups, the perspective is that fishermen could lose the conflict in many countries, particularly in the developed world, with very significant political, financial and cultural losses. While the sector itself may survive because alternative ways of producing 150 million tonnes of food are simply impractical, many fisheries may be closed, starting from the most environmentally "aggressive" ones and in areas where the contribution to food is not essential. Whether this scenario would be favourable to small-scale (more ecological?) or large-scale (more politically effective?) industries is not clear.

(2) A smooth confluence between the two paradigms through improved collaboration between existing institutions or the creation of new, integrated ones. The fisheries stakeholders are put in the position of effectively adopting and implementing a more ecosystem and precautionary approach to development and management. The ecosystem stakeholders realize that in many areas fishing is needed and the fisherman is an "endangered species", essential for the functioning of the coastal communities. Both groups of stakeholders realize that future long-term benefits can operate as incentives today only if the issue of resources allocation is resolved. Both realize and accept that, in the short term, a major problem is in the fact that the cost of change can easily be overwhelming. Both agree that the solution is neither in status quo (fishing as usual) nor in systematic exclusion of fishers through bans or moratoria, and that the challenge is to work out a transition at a pace that is effective enough in reaching the goal and socio-economically acceptable.

Predictions and generalizations as to which scenario will prevail would be risky. While a range of positions exist in both paradigms and both include extremists, option 2 seems to be the one generally favoured by national governance but often looked at reluctantly by a defying sector. It is also the one implicit in most scientific papers with some statement about the future.

It seems indeed that an outright elimination of fisheries as a main use of the aquatic ecosystems is highly unlikely. The ecological footprint of an alternative solution to production of the needed 150 million tonnes of food now coming from the sea would probably be worse and the socio-economic earthquake resulting from the demise of world fisheries unacceptable to society.

A global disappearance of fisheries is unlikely and, as a consequence, a progressive change towards more ecological fisheries will probably happen. Based on the past 50 years' experience, however, the change will happen at different paces in different places and fisheries. Further degradation can therefore be expected in some places while in others the situation will markedly improve. More stocks will collapse. Others will be closed down by management, under public pressure, before this happens. In both cases socio-economic disruption will be very damaging, most often to poor people for which fishing is the only livelihood.

Capture fisheries have sometimes been considered a sunset industry, spiralling down through plummeting resources and profitability as conventional management failed to maintain target stocks at the appropriate level. Will EAF be able to reverse the image? Will EAF succeed in maintaining target stocks and, in addition, the biochemical environment and the habitat? The reply to this question will depend on the circumstances, the political will, the capacity, etc. In any case, EAF appears as a necessary if not sufficient condition to the maintenance of fisheries in the long term.

10

ENVIRONMENTAL IMPACT OF FISHING

The environmental impact of fishing can be divided into issues that involve the availability of fish to be caught, such as overfishing, sustainable fisheries, and fisheries management; and issues that involve the impact of fishing on other elements of the environment, such as by-catch. These conservation issues are part of marine conservation, and are addressed in fisheries science programs. There is a growing gap between how many fish are available to be caught and humanity's desire to catch them, a problem that gets worse as the world population grows. Similar to other environmental issues, there can be conflict between the fishermen who depend on fishing for their livelihoods and fishery scientists who realise that if future fish populations are to be sustainable then some fisheries must reduce or even close.

The journal Science published a four-year study in November 2006, which predicted that, at prevailing trends, the world would run out of wild-caught seafood in 2048. The scientists stated that the decline was a result of overfishing, pollution and other environmental factors that were reducing the population of fisheries at the same time as their ecosystems were being degraded. Yet again the analysis has met criticism as being fundamentally flawed, and many fishery management officials, industry representatives and scientists challenge the findings, although the debate continues. Many countries, such as Tonga, the United States, Australia and New Zealand, and international management bodies have taken steps to appropriately manage marine resources.

EFFECTS ON HABITAT

Some fishing techniques also may cause habitat destruction. Dynamite fishing and cyanide fishing, which are illegal in many places, harm surrounding habitat. Bottom trawling, the practice of pulling a fishing net along the sea bottom behind trawlers, removes around 5 to 25% of an area's seabed life on a single run. A 2005 report of the UN Millennium Project, commissioned by UN Secretary-General Kofi Annan, recommended the elimination of bottom trawling on the high seas by 2006 to protect seamounts and other ecologically sensitive habitats.

In mid October 2006, US President Bush joined other world leaders calling for a moratorium on deep-sea trawling, a practice shown to often have harmful effects on sea habitat and, hence, on fish populations.

Overfishing

Overfishing has also been widely reported due to increases in the volume of fishing hauls to feed a quickly growing number of consumers. This has led to the breakdown of some sea ecosystems and several fishing industries whose catch has been greatly diminished. The extinction of many species has also been reported. According to a Food and Agriculture Organization estimate, over 70% of the world's fish species are either fully exploited or depleted. According to the Secretary General of the 2002 World Summit on Sustainable Development, "Overfishing cannot continue, the depletion of fisheries poses a major threat to the food supply of millions of people."

The cover story of the May 15, 2003 issue of the science journal Nature - with Dr. Ransom A. Myers, an internationally prominent fisheries biologist (Dalhousie University, Halifax, Canada) as the lead author - was devoted to a summary of the scientific information. The story asserted that, as compared with 1950 levels, only a remnant (in some instances, as little as 10%) of all large ocean-fish stocks are left in the seas. These large ocean fish are the species at the top of the food chains (e.g., tuna, cod, among others). This article was subsequently criticized as being fundamentally flawed, although much debate still exists (Walters 2003; Hampton et al. 2005; Maunder et al. 2006; Polacheck 2006;Sibert et al. 2006) and the majority of fisheries scientists now consider the results irrelevant with respect to large pelagics (the open seas).

Ecological disruption

Fishing may disrupt food webs by targeting specific, in-demand species. There might be too much fishing of prey species such as sardines and anchovies, thus reducing the food supply for the predators. It may also cause the increase of prey species when the target fishes are predator species such as salmon and tuna. Fisheries can reduce fish stocks that cetaceans rely on for food.

By-catch

By-catch is the portion of the catch that is not the target species. These are either kept to be sold or discarded. In some instances the discarded portion is known as discards.

Possible remedies

Many governments and intergovernmental bodies have implemented fisheries management policies designed to curb the environmental impact of fishing. Fishing conservation aims to control the human activities that may completely decrease a fish stock or washout an entire aquatic environment. These laws include the quotas on the total catch of particular species in a fishery, effort quotas (e.g., number of days at sea), the limits on the number of vessels allowed in specific areas, and the imposition of seasonal restrictions on fishing..

In 2008 a large scale study of fisheries that used individual transferable quotas and ones that didn't provided strong evidence that individual transferable quotas can help to prevent collapses and restore fisheries that appear to be in decline.

Fish farming has been proposed as a more sustainable alternative to traditional capture of wild fish. However, fish farming has been found to have negative impacts on nearby wild fish. Further, farming of predatory fish like salmon can rely on fish feed that is based on fish meal and oil from wild fish.

The environmental impact of recreational fishing may be alleviated to some extent by catch and release fishing.

Fish Farming

Fish farming is the principal form of aquaculture, while other methods may fall under mariculture. Fish farming involves raising fish commercially in tanks or enclosures, usually for food. A facility that releases juvenile fish into the wild for recreational fishing or to supplement a species' natural numbers is generally referred to as a fish hatchery. Worldwide, the most important fish species used in fish farming are carp, salmon, tilapia and catfish.

There is an increasing demand for fish and fish protein, which has resulted in widespread overfishing in wild fisheries. Fish farming offersfish marketers another source. However, farming carnivorous fish, such as salmon, does not always reduce pressure on wild fisheries, since carnivorous farmed fish are usually fed fishmeal and fish oil extracted from wild forage fish. In this way, the salmon can consume in weight more wild fish than they weigh themselves. The global returns for fish farming recorded by the FAO in 2008 totalled 33.8 milliontonnes worth about $US 60 billion.

Major categories of fish aquaculture

There are two kinds of aquaculture: extensive aquaculture based on local photosynthetical production and intensive aquaculture, in which the fish are fed with external food supply.

EXTENSIVE AQUACULTURE

Limiting for growth here is the available food supply by natural sources, commonly zooplankton feeding on pelagic algae or benthic animals, such ascrustaceans and mollusks. Tilapia species filter feed directly on phytoplankton, which makes higher production possible. The photosynthetic production can be increased by fertilizing the pond water with artificial fertilizer mixtures, such as potash, phosphorus, nitrogen and micro-elements. Because most fish are carnivorous, they occupy a higher place in the trophic chain and therefore only a tiny fraction of primary photosynthetic production (typically 1%) will be converted into harvest-able fish.

Another issue is the risk of algal blooms. When temperatures, nutrient supply and available sunlight are optimal for algal growth, algae multiply their biomass at an exponential rate, eventually leading to an exhaustion of available nutrients and a subsequent die-off. The decaying algal biomass will deplete the oxygen in the pond water because it blocks out the sun and pollutes it with organic and inorganic solutes (such as ammonium ions), which can (and frequently do) lead to massive loss of fish.

An alternate option is to use a wetland system such as that of Veta La Palma.

In order to tap all available food sources in the pond, the aquaculturist will choose fish species which occupy different places in the pond ecosystem, e.g., a filter algae feeder such as tilapia, abenthic feeder such as carp or catfish and a zooplankton feeder (various carps) or submerged weeds feeder such as grass carp.

Despite these limitations significant fish farming industries use these methods. In the Czech Republic thousands of natural and semi-natural ponds are harvested each year for trout and carp. The large ponds around Trebon were built from around 1650 and are still in use.

Intensive aquaculture

In these kinds of systems fish production per unit of surface can be increased at will, as long as sufficient oxygen, fresh water and food are provided. Because of the requirement of sufficient fresh water, a massive water purification system must be integrated in the fish farm. A clever way to achieve this is the combination of hydroponic horticulture and water treatment, see below. The exception to this rule are cages which are placed in a river or sea, which supplements the fish crop with sufficient oxygenated water. Some environmentalists object to this practice.

The cost of inputs per unit of fish weight is higher than in extensive farming, especially because of the high cost of fish feed, which must contain a much higher level of protein (up to 60%) than cattle food and a balanced amino acid composition as well. However, these higher protein level requirements are a consequence of the higher food conversion efficiency (FCR-kg of feed per kg of animal produced) of aquatic animals. Fish like salmon have FCR's in the range of 1.1 kg of feed per kg of salmon whereas chickens are in the 2.5 kg of feed per kg of chicken range. Fish don't have to stand up or keep warm and this eliminates a lot of carbohydrates and fats in the diet, required to provide this energy. This frequently is offset by the lower land costs and the higher productions which can be obtained due to the high level of input control.

Essential here is aeration of the water, as fish need a sufficient oxygen level for growth. This is achieved by bubbling, cascade flow or aqueous oxygen. Catfish, Clarias spp. can breathe atmospheric air and can tolerate much higher levels of pollutants than trout or salmon, which makes aeration and water purification less necessary and makes Clarias species especially suited for intensive fish production. In some Clarias farms about 10% of the water volume can consist of fish biomass.

The risk of infections by parasites like fish lice, fungi (Saprolegnia spp.), intestinal worms (such as nematodes or trematodes), bacteria (e.g.,Yersinia spp., Pseudomonas spp.), and protozoa (such as Dinoflagellates) is similar to animal husbandry, especially at high population densities. However, animal husbandry is a larger and more technologically mature area of human agriculture and better solutions to pathogen problem exist. Intensive aquaculture does have to provide adequate water quality (oxygen, ammonia, nitrite, etc.) levels to minimize stress, which makes the pathogen problem more difficult. This means, intensive aquaculture requires tight monitoring and a high level of expertise of the fish farmer.

Very high intensity recycle aquaculture systems (RAS), where there is control over all the production parameters, are being used for high value species. By recycling the water, very little water is used per unit of production. However, the process does have high capital and operating costs. The higher cost structures mean that RAS is only economical for high value products like broodstock for egg production, fingerlings for net pen aquaculture operations, sturgeon production, research animals and some special niche markets like live fish.

Raising ornamental cold water fish (goldfish or koi), although theoretically much more profitable due to the higher income per weight of fish produced, has never been successfully carried out until very recently. The increased incidences of dangerous viral diseases of koi Carp, together with the high value of the fish has led to initiatives in closed system koi breeding and growing in a number of countries. Today there are a few commercially successful intensive koi growing facilities in the UK, Germany and Israel.

Some producers have adapted their intensive systems in an effort to provide consumers with fish that do not carry dormant forms of viruses and diseases.

Specific types of fish farms

Within intensive and extensive aquaculture methods, there are numerous specific types of fish farms; each has benefits and applications unique to its design.

Cage system

Fish cages are placed in lakes, bayous, ponds, rivers or oceans to contain and protect fish until they can be harvested. The method is also called "off-shore cultivation" when the cages are placed in the sea. They can be constructed of a wide variety of components. Fish are stocked in cages, artificially fed, and harvested when they reach market size. A few advantages of fish farming with cages are that many types of waters can be used (rivers, lakes, filled quarries, etc.), many types of fish can be raised, and fish farming can co-exist with sport fishing and other water uses. Cage farming of fishes in open seas is also gaining popularity. Concerns of disease, poaching, poor water quality, etc., lead some to believe that in general, pond systems are easier to manage and simpler to start. Also, past occurrences of cage-failures leading to escapes, have raised concern regarding the culture of non-native fish species in open-water cages. Even though the cage-industry has made numerous technological advances in cage construction in recent years, the concern for escapes remains valid.

Recently, copper alloys have become important netting materials in aquaculture. Copper alloys are antimicrobial, that is, they destroy bacteria, viruses,fungi, algae, and other microbes. In the marine environment, the antimicrobial/algaecidal properties of copper alloys prevent biofouling, which can briefly be described as the undesirable accumulation, adhesion, and growth of microorganisms, plants, algae, tube worms, barnacles, mollusks, and other organisms.

The resistance of organism growth on copper alloy nets also provides a cleaner and healthier environment for farmed fish to grow and thrive.

In addition to its antifouling benefits, copper netting has strong structural and corrosion-resistant properties in marine environments.

Copper-zinc brass alloys are currently (2011) being deployed in commercial-scale aquaculture operations in Asia, South America and the USA (Hawaii). Extensive research, including demonstrations and trials, are currently being implemented on two other copper alloys: copper-nickel and copper-silicon. Each of these alloy types has an inherent ability to reduce biofouling, cage waste, disease, and the need for antibiotics while simultaneously maintaining water circulation and oxygen requirements. Other types of copper alloys are also being considered for research and development in aquaculture operations.

Irrigation ditch or pond systems

These use irrigation ditches or farm ponds to raise fish. The basic requirement is to have a ditch or pond that retains water, possibly with an above-ground irrigation system (many irrigation systems use buried pipes with headers.) Using this method, one can store one's water allotment in ponds or ditches, usually lined with bentonite clay. In small systems the fish are often fed commercial fish food, and their waste products can help fertilize the fields. In larger ponds, the pond grows water plants and algae as fish food. Some of the most successful ponds grow introduced strains of plants, as well as introduced strains of fish.

Control of water quality is crucial. Fertilizing, clarifying and pH control of the water can increase yields substantially, as long as eutrophication is prevented and oxygen levels stay high. Yields can be low if the fish grow ill from electrolyte stress.

Composite fish culture

The Composite fish culture system is a technology developed in India by the Indian Council of Agricultural Research in the 1970s. In this system both local and imported fish species, a combination of five or six fish species is used in a single fish pond. These species are selected so that they do not compete for food among them having different types of food habitats. As a result the food available in all the parts of the pond is used. Fish used in this system include catla and silver carp which are surface feeders, rohu a column feeder and mrigal and common carp which are bottom feeders. Other fish will also feed on the excreta of the common carp and this helps contribute to the efficiency of the system which in optimal conditions will produce 3000-6000 kg of fish per hectare per year.

Integrated recycling systems

One of the largest problems with freshwater pisciculture is that it can use a million gallons of water per acre (about 1 m of water per m) each year. Extended water purification systems allow for the reuse (recycling) of local water.

The largest-scale pure fish farms use a system derived (admittedly much refined) from the New Alchemy Institute in the 1970s. Basically, large plastic fish tanks are placed in a greenhouse. Ahydroponic bed is placed near, above or between them. When tilapia are raised in the tanks, they are able to eat algae, which naturally grows in the tanks when the tanks are properly fertilized.

The tank water is slowly circulated to the hydroponic beds where the tilapia waste feeds commercial plant crops. Carefully cultured microorganisms in the hydroponic bed convert ammonia to nitrates, and the plants are fertilized by the nitrates and phosphates. Other wastes are strained out by the hydroponic media, which doubles as an aerated pebble-bed filter.

This system, properly tuned, produces more edible protein per unit area than any other. A wide variety of plants can grow well in the hydroponic beds. Most growers concentrate on herbs (e.g.parsley and basil), which command premium prices in small quantities all year long. The most common customers are restaurant wholesalers.

Since the system lives in a greenhouse, it adapts to almost all temperate climates, and may also adapt to tropical climates. The main environmental impact is discharge of water that must be salted to maintain the fishes' electrolyte balance. Current growers use a variety of proprietary tricks to keep fish healthy, reducing their expenses for salt and waste water discharge permits. Some veterinary authorities speculate that ultraviolet ozone disinfectant systems (widely used for ornamental fish) may play a prominent part in keeping the Tilapia healthy with recirculated water.

A number of large, well-capitalized ventures in this area have failed. Managing both the biology and markets is complicated. One future development is the combination of Integrated Recycling systems with Urban Farming as tried in Sweden by the Greenfish initiative.

Classic fry farming

This is also called a "Flow through system" Trout and other sport fish are often raised from eggs to fry or fingerlings and then trucked to streams and released. Normally, the fry are raised in long, shallow concrete tanks, fed with fresh stream water. The fry receive commercial fish food in pellets. While not as efficient as the New Alchemists' method, it is also far simpler, and has been used for many years to stock streams with sport fish. European eel (Anguilla anguilla) aquaculturalists procure a limited supply of glass eels, juvenile stages of the European eel which swim north from the Sargasso Sea breeding grounds, for their farms. The European eel is threatened with extinction because of the excessive catch of glass eels by Spanish fishermen and overfishing of adult eels in, e.g., the Dutch IJsselmeer, Netherlands. As per 2005, no one has managed to breed the European eel in captivity.

Issues

The issue of feeds in fish farming has been a controversial one. Many cultured fishes (tilapia, carp, catfish, many others) require no meat or fish products in their diets. Top-level carnivores (most salmon species) depend on fish feed of which a portion is usually derived from wild caught (anchovies, menhaden, etc.). Vegetable-derived proteins have successfully replaced fish meal in feeds for carnivorous fishes, but vegetable-derived oils have not successfully been incorporated into the diets of carnivores.

Secondly, farmed fish are kept in concentrations never seen in the wild (e.g. 50,000 fish in a 2-acre (8,100 m) area.) with each fish occupying less room than the average bathtub. This can cause several forms of pollution. Packed tightly, fish rub against each other and the sides of their cages, damaging their fins and tails and becoming sickened with various diseases and infections. This also causes stress.

However, fish tend also to be animals that aggregate into large schools at high density. Most successful aquaculture species are schooling species, which do not have social problems at high density. Aquaculturists tend to feel that operating a rearing system above its design capacity or above the social density limit of the fish will result in decreased growth rate and increased FCR (food conversion ratio - kg dry feed/kg of fish produced), which will result in increased cost and risk of health problems along with a decrease in profits. Stressing the animals is not desirable, but the concept of and measurement of stress must be viewed from the perspective of the animal using the scientific method.

Sea lice, particularly Lepeophtheirus salmonis and various Caligus species, including Caligus clemensi and Caligus rogercresseyi, can cause deadly infestations of both farm-grown and wild salmon. Sea lice are ectoparasites which feed on mucus, blood, and skin, and migrate and latch onto the skin of wild salmon during free-swimming, planktonic nauplii and copepodid larval stages, which can persist for several days. Large numbers of highly populated, open-net salmon farms can create exceptionally large concentrations of sea lice; when exposed in river estuaries containing large numbers of open-net farms, many young wild salmon are infected, and do not survive as a result. Adult salmon may survive otherwise critical numbers of sea lice, but small, thin-skinned juvenile salmon migrating to sea are highly vulnerable. On the Pacific coast of Canada, the louse-induced mortality of pink salmon in some regions is commonly over 80%.

A 2008 meta-analysis of available data shows that salmon farming reduces the survival of associated wild salmon populations. This relationship has been shown to hold for Atlantic, steelhead, pink, chum, and coho salmon. The decrease in survival or abundance often exceeds 50 percent.

Diseases and parasites are the most commonly cited reasons for such decreases. Some species of sea lice have been noted to target farmed coho and Atlantic salmon. Such parasites have been shown to have an effect on nearby wild fish. One place that has garnered international media attention is British Columbia's Broughton Archipelago. There, juvenile wild salmon must "run a gauntlet" of large fish farms located off-shore near river outlets before making their way to sea. It is alleged that the farms cause such severe sea lice infestations that one study predicted in 2007 a 99% collapse in the wild salmon population by 2011. This claim, however, has been criticized by numerous scientists who question the correlation between increased fish farming and increases in sea lice infestation among wild salmon.

Because of parasite problems, some aquaculture operators frequently use strong antibiotic drugs to keep the fish alive (but many fish still die prematurely at rates of up to 30 percent). In some cases, these drugs have entered the environment. Additionally, the residual presence of these drugs in human food products has become controversial. Use of antibiotics in food production is thought to increase the prevalence of antibiotic resistance in human diseases. At some facilities, the use of antibiotic drugs in aquaculture has decreased considerably due to vaccinations and other techniques. However, most fish farming operations still use antibiotics, many of which escape into the surrounding environment.

The lice and pathogen problems of the 1990s facilitated the development of current treatment methods for sea lice and pathogens. These developments reduced the stress from parasite/pathogen problems. However, being in an ocean environment, the transfer of disease organisms from the wild fish to the aquaculture fish is an ever-present risk.

The very large number of fish kept long-term in a single location contributes to habitat destruction of the nearby areas. The high concentrations of fish produce a significant amount of condensed faeces, often contaminated with drugs, which again affect local waterways. However, these effects are very local to the actual fish farm site and are minimal to non-measurable in high current sites.

Concern remains that resultant bacterial growth strips the water of oxygen, reducing or killing off the local marine life. Once an area has been so contaminated, the fish farms are moved to new, uncontaminated areas. This practice has angered nearby fishermen.

Other potential problems faced by aquaculturists are the obtaining of various permits and water-use rights, profitability, concerns about invasive species and genetic engineering depending on what species are involved, and interaction with the United Nations Convention on the Law of the Sea.

In regards to genetically modified farmed salmon, concern has been raised over their proven reproductive advantage and how it could potentially decimate local fish populations, if released into the wild. Biologist Rick Howard did a controlled laboratory study where wild fish and GMO fish were allowed to breed. The GMO fish crowded out the wild fish in spawning beds, but the offspring were less likely to survive. The colorant used to make pen-raised salmon appear rosy like their wild cousins has been linked with retinal problems in humans.

Labeling

In 2005, Alaska passed legislation requiring that any genetically altered fish sold in the state be labeled. In 2006, a Consumer Reports investigation revealed that farm-raised salmon is frequently sold as wild.

In 2008, the US National Organic Standards Board allowed farmed fish to be labeled as organic provided less than 25% of their feed came from wild fish. This decision was criticized by the advocacy group Food & Water Watch as "bending the rules" about organic labeling. In the European Union, fish labeling as to species, method of production and origin, has been required since 2002.

Concerns continue over the labeling of salmon as farmed or wild caught, as well as about the humane treatment of farmed fish. The Marine Stewardship Council has established an Eco label to distinguish between farmed and wild caught salmon, while the RSPCA has established the Freedom Food label to indicate humane treatment of farmed salmon as well as other food products.

Indoor fish farming

An alternative to outdoor open ocean cage aquaculture, is through the use of a recirculation aquaculture system (RAS). A RAS is a series of culture tanks and filters where water is continuously recycled and monitored to keep optimal conditions year round. To prevent the deterioration of water quality, the water is treated mechanically through the removal of particulate matter and biologically through the conversion of harmful accumulated chemicals into nontoxic ones.

Other treatments such as UV sterilization, ozonation, and oxygen injection are also used to maintain optimal water quality. Through this system, many of the environmental drawbacks of aquaculture are minimized including escaped fish, water usage, and the introduction of pollutants. The practices also increased feed-use efficiency growth by providing optimum water quality (Timmons et al., 2002; Piedrahita, 2003).

One of the drawbacks to recirculation aquaculture systems is water exchange. However, the rate of water exchange can be reduced through aquaponics, such as the incorporation of hydroponically grown plants (Corpron and Armstrong, 1983) and denitrification (Klas et al., 2006). Both methods reduce the amount of nitrate in the water, and can potentially eliminate the need for water exchanges, closing the aquaculture system from the environment. The amount of interaction between the aquaculture system and the environment can be measured through the cumulative feed burden (CFB kg/M3), which measures the amount of feed that goes into the RAS relative to the amount of water and waste discharged.

From 2011, a team from the University of Waterloo led by Tahbit Chowdhury and Gordon Graff examined vertical RAS aquaculture designs aimed at producing protein-rich fish species.However, because of its high capital and operating costs, RAS has generally been restricted to practices such as broodstock maturation, larval rearing, fingerling production, research animal production, SPF (specific pathogen free) animal production, and caviar and ornamental fish production. As such, research and design work by Chowdhury and Graff remains difficult to implement. Although the use of RAS for other species is considered by many aquaculturalists to be currently impractical, there has been some limited successful implementation of this with high value product such as barramundi, sturgeon and live tilapia in the US eels and catfish in the Netherlands, trout in Denmark and salmon is planned in Scotland and Canada.

Slaughter methods

Tanks saturated with carbon dioxide have been used to make fish unconscious. Then their gills are cut with a knife so that the fish bleed out before they are further processed. This is no longer considered a humane method of slaughter. Methods that induce much less physiological stress are electrical or percussive stunning and this has led to the phasing out of the carbon dioxide slaughter method in Europe.

Inhumane methods

According to T. Håstein of the National Veterinary Institute, "Different methods for slaughter of fish are in place and it is no doubt that many of them may be considered as appalling from an animal welfare point of view." A 2004 report by the EFSA Scientific Panel on Animal Health and Welfare explained: "Many existing commercial killing methods expose fish to substantial suffering over a prolonged period of time. For some species, existing methods, whilst capable of killing fish humanely, are not doing so because operators don't have the knowledge to evaluate them."Following are some of the less humane ways of killing fish.

- Air Asphyxiation. This amounts to suffocation in the open air. The process can take upwards of 15 minutes to induce death, although unconsciousness typically sets in sooner.
- Ice baths / chilling. Farmed fish are sometimes chilled on ice or submerged in near-freezing water. The purpose is to dampen muscle movements by the fish and to delay the onset of post-death decay. However, it does not necessarily reduce sensibility to pain; indeed, the chilling process has been shown to elevate cortisol. In addition, reduced body temperature extends the time before fish lose consciousness.
- CO2 narcosis.
- Exsanguination without stunning. This is a process in which fish are taken up from water, held still, and cut so as to cause bleeding. According to references in Yue, this can leave fish writhing for an average of four minutes, and some catfish still responded to noxious stimuli after more than 15 minutes.
- More humane methods
- Percussive stunning.
- Electric stunning. This can be humane when a proper current, duration, conductivity, and temperature are present. One advantage is that in-water stunning allows fish to be rendered unconscious without stressful handling or displacement. However, improper stunning may not induce insensibility long enough to prevent the fish from enduring exsanguination while conscious. It's unknown whether the optimal stunning parameters that researchers have determined in studies are used by the industry in practice.

FISHERIES IMPACT ON THE ECOSYSTEM

The ecosystems that support fisheries, together with other economic activities, are subject to a number of alterations of significant relevance to their functioning and resilience and to the goods and services they can provide. Because of our imperfect understanding of ecosystem structure and functioning, as well as the inherent difficulty of distinguishing between natural and human-induced changes, the latter are not always perfectly predictable and/or reversible (Christensen et al., 1996). The following sections elaborate briefly on some types of alteration.

Impacts from fisheries on the environment have been abundantly described and reviewed (Dayton et al., 1995; Goñi, 1998; Kaiser et al., 2003;

Gislason, 2003; Agardy, 2000). More specifically, capture fisheries impact target resources. They reduce their abundance, spawning potential and, possibly, population parameters (growth, maturation, etc.). They modify age and size structure, sex ratio, genetics and species composition of the target resources, as well as of their associated and dependent species. When poorly controlled, fisheries develop excessive fishing capacity, leading to overfishing, with major ecosystem, social and economic consequences.

Fishing may also affect ecological processes at very large scale. The overall impact has been described as comparable, in aquatic systems, to that of agriculture on land in terms of the proportion of the system's primary productivity harvested by humans (Pauly and Christensen, 1995). Overfishing transforms an originally stable, mature and efficient ecosystem into one that is immature and stressed. This happens in various ways. By targeting and reducing the abundance of high-value predators, fisheries deeply modify the trophic chain and the flows of biomass (and energy) across the ecosystem (e.g. Pauly, 1979). They can also alter habitats, most notably by destroying and disturbing bottom topography and the associated habitats (e.g. seagrass and algal beds, coral reefs) and benthic communities.

The alteration of the habitat by various human activities may be physical (e.g. by adding artificial structures like artificial reefs, oil rigs, aquaculture installations), mechanical (e.g. through the "ploughing" effect of dredges and trawls), or chemical (e.g. through injection of nutrients, pesticides, heavy metals, drugs, hormones). Fishing may result in changes in productivity of resources (some positive and some negative) and affects associated species. Some aspects of fisheries can have significant and long-lasting effects, e.g. destructive fishing techniques using dynamite or cyanides or inadequate fishing practices (e.g. trawling in the wrong habitat); pollution from fish processing plants; use of ozone-depleting refrigerants; dumping at sea of plastic debris that can entangle marine animals or be swallowed by turtles; loss of fishing gear, possibly leading to ghost fishing; lack of selectivity, affecting associated and dependent species, resulting in wasteful discarding practices, juvenile mortality, added threat to endangered species, etc. Poorly-managed large-scale mariculture can damage coastal wetlands and nearshore ecosystems, often used as nurseries by key capture fishery resources, and contribute to ecosystem contamination with food residues, waste, antibiotics, hormones, diseases and alien species.

Impact on Associated and Dependent Species

The Law of the Sea provides that fisheries management must take care also of associated and dependent species. The impact of fishing on these species has been documented in some areas (Goñi, 1998; Gislason,

2003) but is still frequently unknown or only partly understood. The decline of primary productivity consumers low in the food chain removes important forage species needed higher in the food web, with cascading effects for the ecosystem. Conversely, the removal of top predators such as mammals, tuna or sharks, may release an unusually large abundance of preys at lower levels with cascading and feedback effects on the food chain and species composition (Trites, 2003; Cury et al., 2003). For example, as most sharks and some batoid fishes (angel fishes) are predators located at or near the top of marine food webs, their depletion modifies the intricate trophic interactions of their ecosystems (Pauly and Murphy, 1982; Jackson et al., 2001). The removal of predators through fishing in Kenyan reefs resulted in the expansion of sea urchin population, which apparently led to a decrease in live coral and to loss of topographic complexity, species diversity and fish biomass (McClanahan and Muthiga, 1988). Goñi (1998) reports that the hunting of sea otters (Enhydra lutris) in the Northeast Pacific caused a large-scale expansion of sea urchins, the increased grazing of which caused the decline of the important kelp forest. She also reports that, in the Bering Sea, the expansion of the fisheries on pollock (Theragra chalcogramma) during the 1970s has been considered as a probable cause of the decline of several populations of marine mammals, e.g. sea lions (Eumetopias jubatus) by 76%, seals (Callorhinus ursinus) by 60% and (Phoca vitulina) by 85%, as well as the decline of several seabird populations (Urea algae, U. lomvia, Rissa brevirostris, R. tridactyla). All these non-fish species compete directly with the Pollock fisheries since the target species represent 21-90% of their diet. Impacts can be particularly serious on cartilaginous fish populations, which have a lower productivity and resilience than bony fishes. As a consequence, fisheries targeting shark have a low record of sustainability and some species of skates, sawfish and deep-water dogfish have been virtually extirpated from large regions (Garcia and Majkowski, 1990; Stevenset al., 2000).

A well documented example of direct impact on benthic species is that of modern towed gear (trawls and dredges) which caused, inter alia, long-term changes in abundance and species composition in the Wadden Sea (Goñi, 1998) and Australia (Moran and Stephenson, 2000). The mortality of benthic species associated with or preyed upon by target bottom fish resources resulting from the use of trawls can vary greatly, depending on how the gear is built or rigged (Moran and Stephenson, 2000). For example, the addition of a tickler chain on a commercial beam trawl will allow it to catch more of its bottom-fish target but, at the same time, will detach and uproot more benthic species (as bycatch) (Kaiser et al., 1996 and 1998; Dayton et al., 1995). Likewise, some gear modification can reduce fishing mortality. For example, operating a semi-pelagic trawl 15 cm above the sea bottom has no measurable effect on the benthos community, while the standard demersal trawling dragging on the bottom reduces benthos density by 15.5% (Moran and Stephenson, 2000). Finally, fishing can have significant impact on the genetic diversity of resources and can permanently change populations characteristics (Kenchington, 2003).

Impact on the Environment

Fishing gear can change the living and non-living environment within which the target and other related resources live. Environmental damage may come from the very nature of the fishing technology (e.g. in the use of dynamite or poison) or from the inappropriate use of an otherwise acceptable gear (e.g. using trawls in coral reefs or seagrass beds).

The use of dynamite and other explosives for "blast fishing" is still common in parts of Asia, Africa, Caribbean and South Pacific. A relatively small explosive (beer-bottle size) is capable of destroying a three-metre circular area of stony corals (Goñi, 1998). These practices are generally officially banned by fisheries regulations and laws but often persist because the people involved have little, if any, alternative livelihood.

The impact on the habitat depends on the gear and sediment type. Highly dynamic, soft bottoms (e.g. coarse sand, hydraulic dunes) may suffer limited damage even when exploited by heavy (including hydraulic) dredges. On the contrary, stable, hard, and highly structured habitats (such as coral reefs, seagrass beds, sponge beds) will be easily damaged. One well-documented example is the use of modern towed gear (trawls and dredges) which caused, inter alia, destruction of seagrass beds (Posidonia oceanica) in the Mediterranean and destruction of the oyster (Cassostrea virginica) habitat in Chesapeake Bay (Goñi, 1998). Damage is also related to fishing frequency, gear weight and rigging. Addition of heavy tickler chains to the trawl ground rope increases bottom abrasion and turbidity (Kaiser et al., 1996 and 1998; Dayton et al., 1995) while adding rollers reduces it. The use of sodium cyanide in the Philippines to catch marine tropical fishes for the aquarium trade has led to the destruction of the coral reef habitat and decline of aquarium and food fish.

Poor Selectivity, Bycatch and Discards

Fishing generates bycatch and discards. A first attempt to address the issue at global level was made in the late 1990s by FAO (1997c) and the first estimate of the global extent of the problem (about 27 million tonnes of resources dumped per year) was published by this Organization (Alverson et al., 1994). A recent review of the issue has been undertaken by Cook (2003). Most fishing activities are not selective enough to remove from the ocean only the desired targets and will probably never be. This leads to accidentally catching other species (bycatch), part of which has little or no use (at least in the local context) and will be dumped overboard (as discards) together with the offal from fish processing at sea. The effect is to increase availability of food to scavenger species (including sea birds) and, when concentrated over time, may cause local anoxia of the seabed environment. The resulting amount of organic material may not be negligible. On the North Pacific shelf and upper slope, the amount of offal generated by at-sea production of surimi (a protein extract of fish) is very significant, since the technology used extracts less than 50% of the wet weight from the total catch, the rest being dumped. In the North Sea, 6.5 to 12.5% of the groundfish caught is dumped at sea. Some of this is consumed by sea birds, but a certain amount of offal becomes available to benthic scavengers. The increase in abundance of dogfish (Scyliorhinus canicula) in northern Spain fisheries and that of Raja radiata in Greenland shrimp fisheries has been associated with increased discards. Oxygen depletion due to excess organic loading from discards has been recorded in the New Zealand fisheries for hoki (Macruronus novaezelandie) as well as in the North Eastern Atlantic (Goñi, 1998).

Bycatch mortality also affects many non-fish species which are relevant to the functioning of the overall ecosystem. For example, surface and sub-surface driftnet and long-line fisheries have serious negative effects on populations of sea birds, e.g. albatrosses and petrels in long-line fisheries in the North Pacific and in the Southern Ocean. High seas drift nets have had a considerable impact on sea birds in the northern Pacific as have gillnets in southwest Greenland, eastern Canada and elsewhere.

Gear Loss and Ghost Fishing

Voluntary dumping or loss of fishing gear may lead to ghost fishing. The scale of the impacts of ghost fishing is basically unknown but there are indications that the effects are not negligible (Goñi, 1998). Species affected by discarded gear include not only teleost fish but sea birds, marine mammals and turtles. For example, the incidence of entanglement of marine mammals in floating synthetic debris in the Bering Sea has been related to growing fishing effort and increased use of plastic. Fowler (1987) concludes that entanglement is the principal cause of the current decline in the fur seal population of the Pribilof Islands, accounting for 15% of the mortality of youngsters. As an average, a northern fur seal (Callorhinus ursinus) is expected to encounter 3-25 pieces of net debris along the 800-km yearly migration in the Northeast Pacific. Fish traps, unless made of biodegradable material, contribute to the problem. To illustrate this problem, 31 600 pots were lost in the Bristol Bay crab fishery in a period of two years (Kruse and Kimber, 1993; Goñi, 1998).

FISHERIES VERSUS OTHER IMPACTS

The ocean is used for a wide range of human consumptive and non-consumptive uses providing recreation, food, livelihood, energy and pharmaceuticals. It is also used for activities such as transportation, defence, mining, conservation and scientific research (Richardson, 2003; Rosenberg, 2003). Ample documentation of these uses can be found on the web-based UN Atlas of the Oceans. Non-fishing activities, whether coastal or continental, may have major impacts on the aquatic ecosystems through contamination, habitat modifications and alteration of freshwater flows (FAO, 1995a). The relevance to EAF stems from the fact that most of these impacts will threaten fisheries sustainability, either reducing resource productivity, stability and resilience to fishing, or reducing the quality of the fish as food through contamination.

Through land drainage, sewage, river outflow, wind and rainfall, such economic activities as agriculture, manufacturing or chemical industries, incineration of toxic wastes, human settlements, etc., release excess nutrients (e.g. nitrates, phosphorus) as well as contaminants (e.g. polychlorinated biphenyls (PCBs), mercury, dioxin), radioactive wastes, oil, antifouling paints (tributyl tin), human pathogens (e.g. cholera, salmonella), plastic and other debris. Coastal activities, including human settlements and tourism, often result in conversion and destruction of habitats of high relevance to fisheries such as estuaries or coastal wetlands used by fishery resources as reproduction, nursery or feeding areas, reducing fisheries productivity and resilience. Irrigation and production of hydroenergy reduce freshwater inflows into the oceans, resulting in modification or suppression of seasonal floods (as in the Nile), reducing or eliminating key environmental signals and reducing influx of nutrients. Dams and drying-up of streams, together with mining of gravel and sand and deforestation, modify the quality of habitats and alter or interrupt the reproductive migration processes of many species of marine fish (e.g. salmon, sturgeon). Together, these impacts result in eutrophication, harmful algal blooms (red tides) (for more details on harmful algal blooms, accidental death of animals through entanglement in (or ingestion of) debris; food contamination, human diseases, climate and sea-level changes; UV and temperature changes with impacts on primary productivity and biological habitats (such as coral bleaching) and can lead to total eradication of productive habitats (e.g. in the dried up Aral Sea).

Relative Importance of Fisheries and Other Impacts

Assessing the scale of fisheries effects relative to other impacts can be difficult, because of confounding and interacting combinations with other anthropogenic effects (e.g. pollution, habitat degradation, climate change) and natural variability of environmental factors. In the early 1970s, pollution and habitat degradation originating from land-based activities were considered to be the main factors of fisheries degradation (Stevenson, 1973). More recent publications hold that excessive fishing has become the main destabilizing factor of ecosystems, directly, through removals and associated impacts, as well as indirectly, through the aggravation of eutrophication and subsequent oxygen depletion (Jackson et al., 2001).

While fishing activities are generally fairly well described and their impact on resources have been studied and documented for decades, land-based sources pollution and their impact on the marine ecosystem and on fisheries are very poorly documented. As a consequence, fisheries impacts tend to be the ones more "visible" to the public. Because the fisheries sector is economically and socially weaker than the agricultural or industrial sector, there is a risk that governments and NGOs target it primarily when looking for short-term solutions to the growing environmental problem, raising the issue of intersectoral equity.

Impact on Diadromous Fish

Diadromous fish, using both the inland and marine environments, illustrate the impact of non-fishery activities on fisheries. Habitat alteration, degradation and outright destruction by non-fishery activities are probably the primary cause of extinction in diadromous fishes (Jonsson et al., 1999). For example, among the diadromous species, the sturgeons are particularly threatened, with 13 of the 15 existing species being endangered, and restoration programmes and restocking will not lead to recovery unless the fundamental sources of decline (in addition to overexploitation), such as pollution and habitat degradation, are addressed (Jonsson et al., 1999). Many changes in fish stocks can be seen to result from changing of river runoff or influx of a layer of warm water (Mann, 1993). For example, the collapse of commercial snook fisheries in Florida during the 1950s has been attributed to habitat alteration and reduction of freshwater flow (Zerbi, 1999). Major environmental disturbances responsible for the decline of Salmo salarpopulations include the development of hydroelectric dams, which impede migration to spawning grounds, and acid rain, which degrades quality of nursery grounds. The poor state of Pacific salmon (Oncorhynchus spp.) is also a consequence of the economic development and exploitation of Northwest Pacific ecosystems, including trapping (furs), mining, timber harvesting, grazing, irrigation, dams, municipal and industrial development, pollution and excessive harvest (Lichatowich et al., 1999).

Competition Between Humans and Marine Mammals

Marine mammals are important predators located towards the apex of the oceanic food chain. They have been heavily depleted in the past and are still, in some cases, threatened with extinction. After decades of protection, many populations are again (or still) very abundant, and their feeding requirements place a very significant demand on ocean productivity. Some countries and stakeholder groups are actively promoting the full protection of these species, questioning our capacity to establish a sustainable regime of exploitation. Other countries consider these species as essential resources and as competitors to humans, pointing to the very large quantities of fish consumed by them (Tamura, 2003).

Allocation Implications

Non-fishery impacts on resource abundance result in a reduction of the fishery resources available to humankind. While they may not be perceived as such by the fishery sector, they represent de facto a forced, non-negotiated, non-transparent, and potentially non-sustainable allocation of aquatic space and resources to non-fishing activities, often located far inshore (e.g. sources of pollution). One implication of the EAF is the need to deal with this problem in a more balanced and transparent way.

BIBLIOGRAPHY

- Alverson, D. L., Freeberg, M. H., Murawski, S. A., and Pope, J.G. 1994. A global assessment of fisheries bycatch and discards. FAO Fisheries Technical Paper, 339. FAO, Rome. 233pp.
- Baird, RW and Guenther, TJ (1995). Account of harbour porpoise (Phocoena phocoena) strandings and bycatches along the coast of British Columbia. Rep. Int. Whal. Commn. Special Issue. 16: 159-168.
- Barlow, J & Cameron, GA (2003). Field experiments show that acoustic pingers reduce marine mammal bycatch in the Californian drift gill net fishery. Marine Mammal Science. 19: 265-283.
- Bell J.D., Ratner B.D., Stobutzki I., Oliver J. Addressing the coral reef crisis in developing countries (2006) Ocean and Coastal Management, 49 (12), pp. 976-985.
- Caswell, H, Brault, S, Read, AJ, and TD Smith (1998). Harbour porpoise and fisheries: an uncertain analysis of incidental mortality. Ecological Applications. 8(4): 1226-1238.
- Charles, A. T. 2001. Sustainable Fishery Systems. Blackwell Science, London. 384 pp.
- Cochrane, K.L. 2000. Reconciling sustainability, economic efficiency and equity in fisheries: the one that got away? Fish and Fisheries, 1: 3-21.
- Cooper A. B. (2006) A Guide to Fisheries Stock Assessment From Data to Recommendations" University of New Hampshire.
- Davis K G and Quelch G D (2008) Observer Professionalism Working Group (OPWG) 5th International Fisheries Observer Conference (IFOC) Report. 87pp.
- de Vries, Gerda; Hillen, Thomas; Lewis, Mark; Schonfisch, Birgitt and Muller, Johannes (2006) A Course in Mathematical Biology SIAM. ISBN 978-0-89871-612-2
- Demaster, DJ, Fowler, CW, Perry, SL, and ME Richlen (2001). Predation and competition: the impact of fisheries on marine mammal populations over the next one hundred years. Journal of Mammology. 82: 641-651.
- Donovan, GP. 1994. Developments on issues relating to the incidental catches of cetaceans since 1992 and the UNCED conference. Report of the IWC (Special Issue). 15: 609-613.

- Encyclopædia Britannica Online 25 August 2005 "Population Ecology" article section on Population Cycles
- European Commission. 1994. The New Common Fisheries Policy. Office for Official Publications of the European Communities, Luxembourg. 46pp.
- FAO. 1993. Report of the Expert Consultation on the regulation of fishing effort (fishing mortality). A preparatory meeting for the FAO World Conference on fisheries management and development. Rome, 17-26 January 1983. FAO Fisheries Report R289. FAO, Rome. 34pp.
- FAO. 1995a. Code of Conduct for Responsible Fisheries. FAO, Rome. 41pp.
- FAO. 1995b. Precautionary Approach to Fisheries. Part 1: Guidelines on the Precautionary Approach to Capture Fisheries and Species Introductions. FAO Fisheries Technical Paper 350/1. FAO, Rome. 52pp.
- FAO. 1996. FAO Technical Guidelines for Responsible Fisheries No. 2. Precautionary Approach to Capture Fisheries. FAO, Rome. 54pp.
- FAO. 1996. Precautionary approach to fisheries. Part 1: Guidelines on the precautionary approach to capture fisheries and species introductions. FAO Fisheries Technical Paper 350/1. FAO, Rome.
- FAO. 1997. FAO Technical Guidelines for Responsible Fisheries No. 4: Fisheries Management. FAO, Rome. 82pp.
- FAO. 2000. The State of World Fisheries and Aquaculture. 2000. FAO, Rome.
- Fernø, A. and Olsen, S. (eds.). 1994. Marine Fish Behaviour in Capture and Abundance Estimation. Fishing News Books, Blackwell Science Ltd., Oxford. 221pp.
- Fishing Technology Service (2008) Managing bycatch and reducing discards: Taking it to the next level

INDEX

A

Academy 65-67, 69-70
Account 208, 220
Actions 91, 93
Adaptation 11
Advisory 93
Advocacy 76
Agency 63
Agenda 82
Agreement 213
Alloys 234
Analysis 81, 137
Animals 20, 28, 109
Antibiotic 183-184, 186-187, 238
Aquatic 2, 27, 39, 81, 90, 97, 123-124, 129, 149, 167, 180
Average 57

B

Bacteria 184-185
Basins 86, 138-139
Benthic 20, 87, 192, 243
Biology 41, 68
Biomass 20, 53
Biosphere 46
Blooms 192
Brackish 31
Breathing 30
Breeding 24, 36

C

Canals 142-143
Capacity 126, 217
Carbon 10, 12
Catfish 24-25, 57
Children 57
Chlorine 56
Clarias 233
Coastal 130, 190, 192, 199, 246
Cold-water 85
Commerce 64
Commission 52, 70, 166
Committee 168, 170
Components 123
Consumers 18
Convention 47, 55, 212, 217-219
Copper 234
Crayfish 33
Crocodiles 39

D

Damage 244
Deep-sea 83, 87, 89
Density 107, 237
Diadromous 247
Ditches 25
Diversity 212
Drainage 139, 143

E

Ecological 206, 211
Economic 63, 100, 211

Ecosystems 3, 7, 10, 12, 41, 82-83, 88-90, 123, 126, 200-201, 203, 206-207, 214, 220
Emissions 12
Endangered 59
Endemic 3
Evolution 223-224, 226
Exotic 3

F

Farmers 197
Feeders 20
Feeding 24
Fertiliser 117, 120
Fisheries 7-11, 14-17, 21, 41, 63-66, 68, 70-73, 80-81, 90, 97-102, 123-126, 128, 130-132, 148-159, 165-170, 173, 178-181, 199-204, 207-210, 213-216, 220, 223-224, 226-227, 229-231, 241-243, 245-247, 249-250
Fishing 7-9, 11-17, 19, 21-22, 42, 44, 60, 63, 66, 68, 72, 77, 91, 125, 128, 130, 195, 204, 211, 218, 229-231, 242, 244-245, 250
Flowing 23, 36
Freshwater 1-2, 6, 97, 153
Fungus 4

G

General 64
Global 11, 44, 81, 103, 111, 149, 151
Glochidia 6
Government 95
Growth 105-109, 113-114, 116-117

H

Habitats 2, 86, 212, 246
Harvesting 144
Health vi, 183-184
Holding 134, 139-140
Hotspot 82-83
Hydroponic 236

I

Indicators 174
Industrial 129
Injection 140
Inland 97-98, 100, 102-103, 119, 123, 126, 128, 131-132, 148-159, 161, 165-171, 180-181
Inspector 64
Irrigation 235

K

Kingdom 42

L

Length 145, 177

M

Mammal 249
Management vi, 14, 45, 63-64, 81, 97-99, 123, 125-127, 130-132, 165, 167, 171, 174, 178-180, 199-210, 212-218, 221, 223-225, 229
Manures 27, 117
Marine 7-8, 12-15, 19-20, 32, 41-44, 46-49, 51-52, 54-55, 63, 65, 67-71, 77, 89, 149, 181, 184, 187, 204, 210, 212-215, 247, 249
Markets 173
Mechanisms 130
Methane 88, 104, 109
Minamata 58
Molluscs 37
Murrel 30
Mussels 6

N

Networks 49
Nitrogen 193
Nursery 138

O

Operations 95, 197
Ornamental 29
Oxidation 31
Oxygen 25, 140, 233
Oysters 20

P

Parasites 4
Plants 56
Policy 76, 81, 174
Pollution 13, 54-55, 129, 221, 247
Population 3, 104, 107
Prawns 29, 31, 33-34
Predators 243
Production 11, 72, 97-100, 102, 108, 120, 135, 139, 148-149, 151-158, 160-161, 165, 191, 232-233, 240
Protection 42, 45, 144, 212

Q

Quality 97, 128, 137, 239
Quantities 117

R

Rainbow 3
Reservoirs 102, 111, 113, 119, 130
Resistance 184
Resolution 121
Rivers 36, 55, 98, 129-130

S

Salinity 1
Salmon 7, 25, 28, 57, 72, 76, 78, 186-191, 193-198, 231, 233, 237-239
Salmonid 190-191, 195
Seafood 15, 73-75
Seagrass 244
Sediments 185, 192
Sewage 30-31
Shellfish 57
Shrimps 33-34
Species 1-5, 8-9, 15, 17, 26, 34, 41, 59, 63, 88, 124-125, 127-128, 132, 159-163, 175-177, 192-193, 199, 202, 212, 219-220, 230-231, 235, 237-238, 240, 242-243, 245
Spillways 147
Surveys 173
Swamps 30, 119
Systems 126-127, 235

T

Taxonomic 159
Temperate 23, 28
Tilapia 24, 105-109, 114, 116-117, 162, 236
Tissue 56
Trawling 16, 22, 77
Trematodes 5

V

Vaccines 188
Vessel 92
Vulnerable 22

W

Waters 6, 45, 106-107, 114, 119, 153
Wetlands 98